Irrigation potential in Africa

A basin approach

FAO LAND AND WATER BULLETIN

4

FAO Land and Water Development Division

Food and Agriculture Organization of the United Nations

Rome, 1997

The designations employed and the presentation of material in this publication do not imply the expression of any opinion whatsoever on the part of the Food and Agriculture Organization of the United Nations concerning the legal status of any country, territory, city or area or of its authorities, or concerning the delimitation of its frontiers or boundaries.

M-54
ISBN 92-5-103966-6

Foreword

Irrigation is viewed as a key factor in progress towards achieving food security in Africa. While nearly 40% of the world's agricultural production comes from irrigated land, the figure for sub-Saharan Africa is only 10%. For most countries of the region, including some poorly endowed with water, only a small part of the available water is withdrawn for use, owing to the state of underdevelopment of water management infrastructure.

Assessment of irrigation potential is of prime importance for planning of sustainable food production in the continent. Considerable information on irrigation potential exists in the African countries, but because of the large numbers of international rivers the regional dimension of the African water resources requires an approach ensuring consistency both within the country and among countries within each river basin.

The present study combines a review of existing information on irrigation potential by country with an approach using a geographic information system to assess land and water availability for irrigation on the basis of river basins. The ever-present environmental issues related to water management highlight some of the major challenges to irrigation development on the continent.

The results of this study and the methodology developed in the report should be useful to researchers and planners at national and regional levels for work aiming at sustainable water resources development in Africa.

Acknowledgements

This report was prepared by Karen Frenken, in collaboration with Jean-Marc Faurès, of the Land and Water Development Division of FAO. It is based on the results of a study performed with the assistance of several contributors: Freddy Nachtergaele for the soil and terrain suitability, Luc Verelst for the irrigation water requirements, Beatrice Crescenzi-Teodori for the geographic information system, and Ahmed Belfouzi for the environmental aspects of irrigation development.

The report was reviewed by Robert Brinkman, Arumugam Kandiah and Jacques Chabloz, who provided useful comments and suggestions for improvement.

The final maps were prepared by Marco Tagliaferri and Raja Refk, editing was done by Julian Plummer and the report was prepared for printing by Chrissi Smith-Redfern.

Contents

List of figures

List of maps

List of tables

Chapter 1

Introduction

There is growing concern about food security in Africa and especially in sub-Saharan Africa. While the aggregate global food supply/demand picture is relatively good, there will be a worsening in food security in sub-Saharan Africa and cereal imports are projected to triple between 1990 and 2020; imports for which the region will not be able to pay. Although the food situation is less severe in north Africa, projections here too indicate increasing cereal imports to 2020 [24].

Africa, with the exception of the Congo/Zaire river basin, is the driest continent (apart from Australia) and suffers from the most unstable rainfall regime. Droughts are frequent in most African countries and each year more people are at risk from the effects of inevitable droughts of greater or lesser severity. Furthermore, Africa's water resources are relatively less developed than those in other regions.

Agricultural productivity per caput in sub-Saharan Africa has not kept pace with population increase, and the region is now in a worse position nutritionally than it was 30 years ago: food production has achieved a growth of about 2.5% per year, while population has risen at a rate of over 3% per year. In the past, additional food in Africa came from increase in the area cultivated, but as good land becomes less available, the region will be forced to increase yields. Both rainfed and irrigated agriculture will need to be intensified, but irrigated agriculture has a higher potential for intensification.

Global estimates indicate that irrigated agriculture produces nearly 40% of food and agriculture commodities on 17% of agricultural land. At present in Africa, about 12.2 million hectares benefit from irrigation[1], which is equal to only about 8.5% of the cultivated land [21a]. In sub-Saharan Africa, only about 10% of the agricultural production comes from irrigated land. Trends in irrigated land expansion over the last 30 years show that, on average, irrigation in Africa increased at a rate of 1.2% per year. However, this rate began to fall in the mid-1980s and is now below 1% per year, but varies widely from country to country [8].

While it is true that there still exists considerable potential for the future expansion of irrigation, it is also true that water is growing scarcer in those regions where the need for irrigation is greatest. Over half of the total water withdrawal takes place in the northern, drier part of Africa. Moreover, in this part the withdrawal for domestic and industrial uses will grow

---

[1] This includes irrigation schemes with full or partial control (11.5 million ha), spate irrigation (0.5 million ha) and wetland and inland valley bottoms, that are equipped for water control (0.2 million ha). Another 2 million hectares benefit from other minor kinds of water management, mainly flood recession cropping (1 million ha) and wetland cropping (1 million ha). In addition, water harvesting methods are becoming more widespread.

fastest, though it will also grow in sub-Saharan Africa in the coming years, as a result of the rapid urbanization.

To enable careful planning of the development of the water resources, especially for agriculture, which is by far the largest water user, a good knowledge of the irrigation potential for the African continent is necessary. This is the subject of the present study.

In the past, several studies have already attempted to assess the irrigation potential for Africa.

In 1987 FAO conducted a study to assess the land and water resources potential for irrigation for Africa on the basis of river basins and countries [20]. It was one of the first GIS-based studies of its kind at continental level. It proposed a natural resources based approach to assessing irrigation potential. Its main limitation was in the sensitivity of the criteria for defining land suitability for irrigation and in the water allocation scenarios needed for the computation of the potential.

In 1995 another study was conducted by FAO as part of the AQUASTAT programme, which is a programme of collection of secondary information on water resources and irrigation by country. A survey was carried out for all African countries, in which information on irrigation potential was systematically collected from master plans and sectoral studies [21a]. Such an approach integrates many more considerations than a simple physical approach to assessing irrigation potential. However, it cannot account for the possible double counting of water resources shared by several countries.

The present study has taken the above limitations into consideration. It concentrates mainly on a quantitative assessment based on physical criteria (land and water), but relies heavily on information collected from the countries. A river basin approach has been used to ensure consistency at river basin level. Where country information was unavailable or incomplete, potential was assessed on the basis of available information on land and water resources at regional and continental level. The FAO Geographic Information System (GIS) facilities were extensively used for this purpose.

A physical approach to irrigation potential must be understood as setting the global limit for irrigation development. Future developments will be dictated by a whole set of factors, including political choices, investment capacity, technological improvement and environmental requirements.

Chapter 2 of this report describes the methodology and data used for the assessment of the irrigation potential. Chapters 3 to 7 refer to a series of detailed studies conducted in the framework of this study. Chapter 3 summarizes the assessment of the soil and terrain suitability for irrigation. Chapter 4 gives a brief review of the African water resources [21]. The computation of irrigation water requirements is summarized in Chapter 5. The main component of the present study, the review of existing information on irrigation potential by basin and country and its cross-checking with the results of the studies of Chapters 3, 4 and 5, is summarized in Chapter 6. Chapter 7 summarizes some environmental considerations in the development of irrigation, though without presuming to be exhaustive on these complex issues. The general results and conclusions are presented in Chapter 8. Finally, a list of the main sources of information is presented by country.

Chapter 2

Methodology and data used

This chapter gives an overview of the methodology used for assessing the irrigation potential and of the different steps followed (Figure 1).

DEFINITION OF IRRIGATION POTENTIAL

This study refers to irrigation as the process by which water is diverted from a river or pumped from a well and used for the purpose of agricultural production. Areas under irrigation thus include areas equipped for full and partial control irrigation, spate irrigation areas, equipped wetland and inland valley bottoms (including fadamas), irrespective of their size or management type. It does not consider techniques related to on-farm water conservation like water harvesting.

The area which can potentially be irrigated depends on the physical resources 'soil' and 'water', combined with the irrigation water requirements as determined by the cropping patterns and climate. In this study it is called 'physical irrigation potential'. However, environmental and socio-economic constraints also have to be taken into consideration in order to guarantee a sustainable use of the available physical resources. This means that in most cases the possibilities for irrigation development would be less than the physical irrigation potential.

DEFINITION OF THE BASIC UNITS

Planning for water use can only be carried out on the basis of river basins. On the other hand, land use is usually computed or planned according to political boundaries. These two divisions of the continent were therefore combined to obtain the basic units used in this study.

First, the African continent was divided into 24 major hydrological units or basin groups, the ones defined in the previous study [20], and classified according to four main categories (Figure 2):

- 8 major river basins, draining to the sea: Senegal River, Niger River, Nile, Shebelli-Juba, Congo/Zaire River, Zambezi, Limpopo and Orange;
- 9 coastal regions grouping several small rivers, draining to the sea: Mediterranean, North-West, West, West Central, South-West, South Atlantic, Indian Ocean, East Central and North-East;
- 5 regions grouping several endorheic drainage basins: Lake Chad, Rift Valley, Okavango, South Interior and North Interior;

- 2 units grouping the islands: one unit is Madagascar and the other unit groups the islands of Cape Verde, Comoros, Mauritius, São Tome and Principe and Seychelles.

The last three categories group several small, independent drainage basins in order to limit the study to a workable number of units.

These 24 major hydrological units were combined with the 53 African countries (in GIS) to obtain 136 land units. These units form the basis of all computation and of the information gathered and analysed in this study and are referred to as 'basic units' (Tables 1 and 2).

IDENTIFICATION OF THE PHYSICAL RESOURCES

Land resources

Criteria were established to determine the soil and terrain suitability for irrigation on the basis of the information from the FAO-UNESCO soil map of the world.

The type of irrigation considered was surface irrigation. Introducing sprinkler irrigation or micro-irrigation on a large scale would require a revision of several of these criteria, probably leading to an increase in land suitable for irrigation.

Water resources

In 1995 FAO has conducted a review of the annual renewable water resources of the African countries [21]. These figures were used as a basis for the present study and completed with more-detailed information on the variation in water discharges in space and time. This information was compared with surface runoff estimates, calculated for each of the 136 basic units (in GIS) and based on the surface runoff map of Africa [28].

All calculations were based on renewable water resources, and mainly on surface water resources, except for arid countries where renewable groundwater already plays an important role in irrigation development. Non renewable groundwater resources (fossil water) were not taken into consideration. For arid countries, this may result in a relatively low irrigation potential, sometimes even lower than the area already under irrigation.

Irrigation water requirements

Assessment of the irrigation potential, based on soil and water resources, can only be done by simultaneously assessing the irrigation water requirements, which in turn depend on the cropping pattern and climate (rainfall and potential evapotranspiration). For this reason, irrigation cropping pattern zones were defined for current and potential scenarios and water requirements were computed using the FAO CROPWAT model [3]. The climate data used were the ones available on the FAOCLIM cd-rom [7]. From the resulting net water requirements, figures of gross irrigation water requirements (including water losses) were computed for each of the 136 basic units. These figures were then compared with those available from individual country studies.

REVIEW OF EXISTING INFORMATION ON IRRIGATION POTENTIAL

The main component of the present study is the review of existing information on irrigation potential, mainly based on physical criteria, though other criteria are sometimes implicitly included as well, as explained in Chapter 6.

In the framework of the AQUASTAT programme, a library has been created containing all kinds of information related to irrigation and indexed by country: national water master plans, agricultural and/or irrigation sector reviews, project documents, country studies, statistics, etc. All these documents were reviewed for the present study.

In addition, more in-depth research on irrigation potential at country, basin and regional level was conducted in the various FAO libraries and connected information systems.

Two river basins, the Nile and the Niger, were subjected to detailed study, while in view of time constraints the other basins and regions were studied at a more global level. To the extent possible, information was collected for each of the 136 basic units. Where it was impossible to have exact figures at these levels, interpolations and/or estimates were made. All the information gathered from the literature was systematically cross-checked with the results of the studies of Chapters 3, 4 and 5.

This study concentrates on water use for agricultural purposes. Where national water master plans exist, water demand by other sectors (domestic, livestock, industrial, hydropower, navigation, etc.) was taken into consideration in assessing water availability for agricultural purposes. Especially in drier regions, competition for water may arise among the different sectors. In general, the quantity of water available for agriculture is the difference between the total quantity of water available and the water demands of other sectors.

ENVIRONMENTAL CONSIDERATIONS

The chief concern of the present study is the physical potential. It is impossible to integrate complex issues, like economic, political, social and environmental aspects into a purely quantitative assessment exercise. Nonetheless these issues are critical to a holistic vision of irrigation potential at continental level. The impact of the choice of one or the other land use, for instance, could radically modify the assessment of land which could be allocated for irrigation. A qualitative assessment of the environmental aspects is presented, highlighting the most relevant issues concerning irrigation development, though without presuming to be exhaustive.

INTERPRETATION OF THE RESULTS

After collating all the information, the figures resulting from the country studies were analysed and compared with the figures resulting from the basin studies in order to develop regional tables. The importance of the issue of water sharing between countries emerges clearly from this confrontation.

FIGURE 1
Assessment of irrigation potential

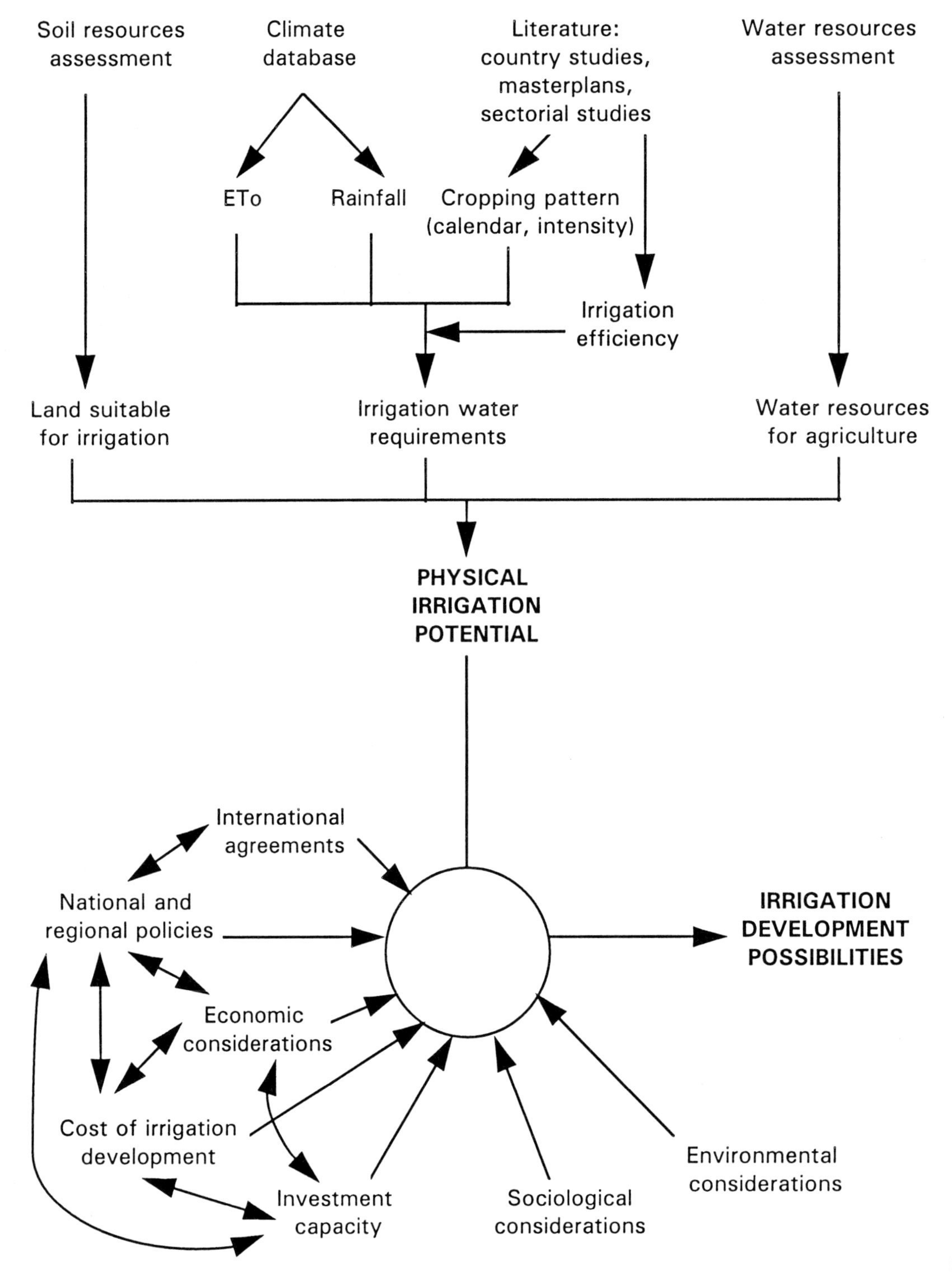

FIGURE 2
Major basin groups of Africa

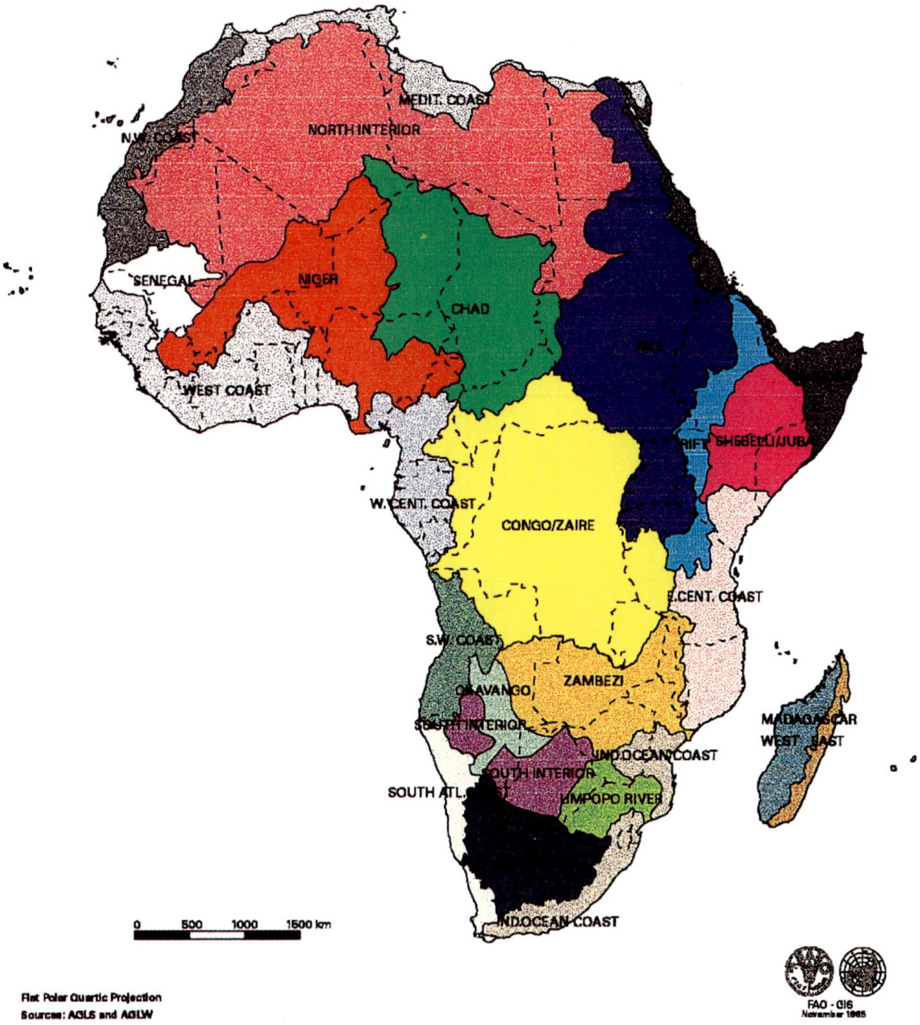

TABLE 1
Areas of the 137 basic units (in km^2)

Basin / Country	Senegal (1)	Niger (2)	Lake Chad (3)	Nile (4)	Rift (5)	Sheb-Juba (6)	Congo/Zaire (7)	Zambezi (8)	Okavango (9)	Limpopo (10)	Orange (11)	South Int. (12)	North Int. (13)	Med.Coast (14)	N.W.Coast (15)	W.Coast (16)	W.C.Coast (17)	S.W.Coast (18)	S.A.Coast (19)	I.O.Coast (20)	E.C.Coast (21)	N.E.Coast (22)	Madagascar (23)	Islands (24)	Area of country	As % of Africa	Basin / Country
Algeria		193 449	93 451										1 944 795	133 327	16 718										2 381 740	7.86%	Algeria
Angola							285 395	235 423	166 963			53 118					7 150	498 651							1 246 700	4.12%	Angola
Benin		46 384														66 236									112 620	0.37%	Benin
Botswana								12 401	49 431	80 118	71 000	368 780													581 730	1.92%	Botswana
Burkina Faso		76 621														197 379									274 000	0.90%	Burkina Faso
Burundi				13 260			14 574																		27 834	0.09%	Burundi
Cameroon		89 249	50 775				96 395										239 021								475 440	1.57%	Cameroon
Cape Verde																								4 030	4 030	0.01%	Cape Verde
Centr. Afr. Rep.			219 410				403 570																		622 980	2.06%	Centr. Afr. Rep.
Chad		20 339	1 046 196										217 465												1 284 000	4.24%	Chad
Comoros																								1 861	1 861	0.01%	Comoros
Congo							246 977										95 023								342 000	1.13%	Congo
Cote d'Ivoire		23 770														298 692									322 462	1.06%	Cote d'Ivoire
Djibouti					12 800																	10 400			23 200	0.08%	Djibouti
Egypt				326 751									520 881	65 568								88 250			1 001 450	3.31%	Egypt
Equat. Guinea																	28 050								28 050	0.09%	Equat. Guinea
Eritrea				24 921	8 605																	88 364			121 890	0.40%	Eritrea
Ethiopia				365 117	310 981	373 739																50 173			1 100 010	3.63%	Ethiopia
Gabon																	267 670								267 670	0.88%	Gabon
Gambia																11 300									11 300	0.04%	Gambia
Ghana																238 540									238 540	0.79%	Ghana
Guinea	29 475	96 880														119 502									245 857	0.81%	Guinea
Guinea Bissau																36 120									36 120	0.12%	Guinea Bissau
Kenya				46 229	130 452	210 226															193 463				580 370	1.92%	Kenya
Lesotho											30 350														30 350	0.10%	Lesotho
Liberia																97 750									97 750	0.32%	Liberia
Libya													1 472 372	287 168											1 759 540	5.81%	Libya
Madagascar																							587 040		587 040	1.94%	Madagascar
Malawi								108 360													10 120				118 480	0.39%	Malawi
Mali	139 098	578 850											512 746			9 496									1 240 190	4.09%	Mali
Mauritania	242 742												578 393		204 385										1 025 520	3.39%	Mauritania
Mauritius																								2 040	2 040	0.01%	Mauritius
Mor.+W.Sah.													154 682	108 300	449 518										712 500	2.35%	Mor.+W.Sah.a
Mozambique								162 004		84 981										185 726	368 879				801 590	2.65%	Mozambique
Namibia								17 426	106 798		219 249	199 718						17 549	264 160						824 900	2.72%	Namibia
Niger		564 211	691 473										11 316												1 267 000	4.18%	Niger
Nigeria		584 193	179 282													101 802	58 493								923 770	3.05%	Nigeria
Rwanda				19 876			6 464																		26 340	0.09%	Rwanda
Sao Tome & Pr.																								960	960	0.00%	Sao Tome & Pr.
Senegal	71 866															124 854									196 720	0.65%	Senegal
Seychelles																								455	455	0.00%	Seychelles
Sierra Leone																71 740									71 740	0.24%	Sierra Leone
Somalia						226 462															19 133	392 065			637 660	2.11%	Somalia
South Africa										185 298	575 769								101 325	358 648					1 221 040	4.03%	South Africa
Sudan			101 048	1 978 506	16 441								313 365									96 450			2 505 810	8.27%	Sudan
Swaziland																				17 364					17 364	0.06%	Swaziland
Tanzania				84 200	153 800		244 593	27 840													434 657				945 090	3.12%	Tanzania
Togo																56 785									56 785	0.19%	Togo
Tunisia													78 448	85 162											163 610	0.54%	Tunisia
Uganda				231 366	4 514																				235 880	0.78%	Uganda
Zaire				22 143			2 313 350										9 367								2 344 860	7.74%	Zaire
Zambia							177 735	574 875																	752 610	2.48%	Zambia
Zimbabwe								213 036		51 467		24 210								102 047					390 760	1.29%	Zimbabwe
Area of basin	483 181	2 273 946	2 381 635	3 112 369	637 593	810 427	3 789 053	1 351 365	323 192	401 864	896 368	645 826	5 804 463	679 525	670 621	1 430 196	704 774	516 200	365 485	663 785	1 026 252	725 702	587 040	9 346	30 290 208	100.00%	TOTALS
As % of Africa	1.60%	7.51%	7.86%	10.28%	2.10%	2.68%	12.51%	4.46%	1.07%	1.33%	2.96%	2.13%	19.16%	2.24%	2.21%	4.72%	2.33%	1.70%	1.21%	2.19%	3.39%	2.40%	1.94%	0.03%	100.00%		
Basin	Senegal	Niger	Lake Chad	Nile	Rift	Sheb-Juba	Congo/Zaire	Zambezi	Okavango	Limpopo	Orange	South Int.	North Int.	Med.Coast	N.W.Coast	W.Coast	W.C.Coast	S.W.Coast	S.A.Coast	I.O.Coast	E.C.Coast	N.E.Coast	Madagascar	Islands	TOTALS		

TABLE 2A
Extent of countries within major basin group (in %)

Basin / Country	Senegal (1)	Niger (2)	Lake Chad (3)	Nile (4)	Rift (5)	Sheb-Juba (6)	Congo/Zaire (7)	Zambezi (8)	Okavango (9)	Limpopo (10)	Orange (11)	South Int. (12)	North Int. (13)	Med.Coast (14)	N.W.Coast (15)	W.Coast (16)	W.C.Coast (17)	S.W.Coast (18)	S.A.Coast (19)	I.O.Coast (20)	E.C.Coast (21)	N.E.Coast (22)	Madagascar (23)	Islands (24)	Country as % of Africa	Basin / Country
Algeria		8.5	3.9										33.5	19.6	2.5										7.86	Algeria
Angola							7.5	17.4	51.7			8.3					1.0	96.6							4.12	Angola
Benin		2.0														4.6									0.37	Benin
Botswana								0.9	15.3	19.9	7.9	57.1													1.92	Botswana
Burkina Faso		3.4														13.8									0.90	Burkina Faso
Burundi				0.4			0.4																		0.09	Burundi
Cameroon		3.9	2.2				2.5										33.9								1.57	Cameroon
Cape Verde																								43.1	0.01	Cape Verde
Centr. Afr. Rep.			9.2				10.7																		2.06	Centr. Afr. Rep.
Chad		0.9	43.9										3.7												4.24	Chad
Comoros																								19.9	0.01	Comoros
Congo							6.5										13.5								1.13	Congo
Cote d'Ivoire		1.0														20.9									1.06	Cote d'Ivoire
Djibouti					2.0																	1.4			0.08	Djibouti
Egypt				10.5									9.0	9.7								12.2			3.31	Egypt
Equat. Guinea																	4.0								0.09	Equat. Guinea
Eritrea				0.8	1.3																	12.2			0.40	Eritrea
Ethiopia				11.8	48.8	46.1																6.9			3.63	Ethiopia
Gabon																	38.0								0.88	Gabon
Gambia																0.8									0.04	Gambia
Ghana																16.7									0.79	Ghana
Guinea	6.1	4.3														8.4									0.81	Guinea
Guinea Bissau																2.5									0.12	Guinea Bissau
Kenya				1.5	20.5	26.0															18.9				1.92	Kenya
Lesotho											3.4														0.10	Lesotho
Liberia																6.8									0.32	Liberia
Libya													25.4	42.3											5.81	Libya
Madagascar																							100.0		1.94	Madagascar
Malawi								8.0													1.0				0.39	Malawi
Mali	28.8	25.5											8.8			0.7									4.09	Mali
Mauritania	50.2												9.9		30.5										3.39	Mauritania
Mauritius																								21.8	0.01	Mauritius
Mor.+W.Sah.													2.7	15.9	67.0										2.35	Mor.+W.Sahara
Mozambique								12.0		21.1										28.0	35.9				2.65	Mozambique
Namibia								1.3	33.0		24.5	30.9						3.4	72.3						2.72	Namibia
Niger		24.8	29.0										0.2												4.18	Niger
Nigeria		25.7	7.5													7.1	8.3								3.05	Nigeria
Rwanda				0.6			0.2																		0.09	Rwanda
Sao Tome & Pr.																								10.3	0.00	Sao Tome & Pr.
Senegal	14.9															8.7									0.65	Senegal
Seychelles																								4.9	0.00	Seychelles
Sierra Leone																5.0									0.24	Sierra Leone
Somalia						27.9															1.9	54.0			2.11	Somalia
South Africa										46.2	64.2								27.7	54.0					4.03	South Africa
Sudan			4.3	63.6	2.6								5.4									13.3			8.27	Sudan
Swaziland																				2.6					0.06	Swaziland
Tanzania				2.7	24.1		6.5	2.1													42.3				3.12	Tanzania
Togo																4.0									0.19	Togo
Tunisia													1.4	12.5											0.54	Tunisia
Uganda				7.4	0.7																				0.78	Uganda
Zaire				0.7			61.0										1.3								7.74	Zaire
Zambia							4.7	42.5																	2.48	Zambia
Zimbabwe								15.8		12.8		3.7								15.4					1.29	Zimbabwe
Total (%)	100.0	100.0	100.0	100.0	100.0	100.0	100.0	100.0	100.0	100.0	100.0	100.0	100.0	100.0	100.0	100.0	100.0	100.0	100.0	100.0	100.0	100.0	100.0	100.0	100.00	Total (%)
Basin	Senegal	Niger	Lake Chad	Nile	Rift	Sheb-Juba	Congo/Zaire	Zambezi	Okavango	Limpopo	Orange	South Int.	North Int.	Med.Coast	N.W.Coast	W.Coast	W.C.Coast	S.W.Coast	S.A.Coast	I.O.Coast	E.C.Coast	N.E.Coast	Madagascar	Islands		

TABLE 2B
Extent of major basin groups within country (in %)

Basin / Country	Senegal (1)	Niger (2)	Lake Chad (3)	Nile (4)	Rift (5)	Sheb-Juba (6)	Congo/Zaire (7)	Zambezi (8)	Okavango (9)	Limpopo (10)	Orange (11)	South Int. (12)	North Int. (13)	Med.Coast (14)	N.W.Coast (15)	W.Coast (16)	W.C.Coast (17)	S.W.Coast (18)	S.A.Coast (19)	I.O.Coast (20)	E.C.Coast (21)	N.E.Coast (22)	Madagascar (23)	Islands (24)	Total (%)	Basin / Country
Algeria		8.1	3.9										81.7	5.6	0.7										100.0	Algeria
Angola							22.8	18.9	13.4			4.3					0.6	40.0							100.0	Angola
Benin		41.2														58.8									100.0	Benin
Botswana								2.1	8.5	13.8	12.2	63.4													100.0	Botswana
Burkina Faso		28.0														72.0									100.0	Burkina Faso
Burundi				47.6			52.4																		100.0	Burundi
Cameroon		18.8	10.6				20.3										50.3								100.0	Cameroon
Cape Verde																								100.0	100.0	Cape Verde
Central African Rep.			35.2				64.8																		100.0	Central African Rep.
Chad		1.6	81.5										16.9												100.0	Chad
Comoros																								100.0	100.0	Comoros
Congo							72.2										27.8								100.0	Congo
Cote d'Ivoire		7.4														92.6									100.0	Cote d'Ivoire
Djibouti					55.2																	44.8			100.0	Djibouti
Egypt				32.6									52.1	6.5								8.8			100.0	Egypt
Equatorial Guinea																	100.0								100.0	Equatorial Guinea
Eritrea				20.4	7.1																	72.5			100.0	Eritrea
Ethiopia				33.1	28.3	34.0																4.6			100.0	Ethiopia
Gabon																	100.0								100.0	Gabon
Gambia																100.0									100.0	Gambia
Ghana																100.0									100.0	Ghana
Guinea	12.0	39.4														48.6									100.0	Guinea
Guinea Bissau																100.0									100.0	Guinea Bissau
Kenya				8.0	22.5	36.2															33.3				100.0	Kenya
Lesotho											100.0														100.0	Lesotho
Liberia																100.0									100.0	Liberia
Libya													83.7	16.3											100.0	Libya
Madagascar																							100.0		100.0	Madagascar
Malawi								91.5													8.5				100.0	Malawi
Mali	11.2	46.7											41.3			0.8									100.0	Mali
Mauritania	23.7												56.4		19.9										100.0	Mauritania
Mauritius																								100.0	100.0	Mauritius
Morocco +W.Sahara													21.7	15.2	63.1										100.0	Morocco +W.Sahara
Mozambique								20.2		10.6										23.2	46.0				100.0	Mozambique
Namibia								2.1	12.9		26.7	24.2						2.1	32.0						100.0	Namibia
Niger		44.5	54.6										0.9												100.0	Niger
Nigeria		63.2	19.5													11.0	6.3								100.0	Nigeria
Rwanda				75.5			24.5																		100.0	Rwanda
Sao Tome & Pr.																								100.0	100.0	Sao Tome & Pr.
Senegal	36.5															63.5									100.0	Senegal
Seychelles																								100.0	100.0	Seychelles
Sierra Leone																100.0									100.0	Sierra Leone
Somalia						35.5															3.0	61.5			100.0	Somalia
South Africa										15.2	47.2								8.3	29.3					100.0	South Africa
Sudan			4.0	79.0	0.7								12.5									3.8			100.0	Sudan
Swaziland																				100.0					100.0	Swaziland
Tanzania				8.9	16.3		25.9	2.9													46.0				100.0	Tanzania
Togo																100.0									100.0	Togo
Tunisia													47.9	52.1											100.0	Tunisia
Uganda				98.1	1.9																				100.0	Uganda
Zaire				0.9			98.7										0.4								100.0	Zaire
Zambia							23.6	76.4																	100.0	Zambia
Zimbabwe								54.5		13.2		6.2								26.1					100.0	Zimbabwe
Basin as % of Africa	1.60	7.51	7.86	10.28	2.10	2.68	12.51	4.46	1.07	1.33	2.96	2.13	19.16	2.24	2.21	4.72	2.33	1.70	1.21	2.19	3.39	2.40	1.94	0.03	100.0	Basin as % of Africa
Basin	Senegal	Niger	Lake Chad	Nile	Rift	Sheb-Juba	Congo/Zaire	Zambezi	Okavango	Limpopo	Orange	South Int.	North Int.	Med.Coast	N.W.Coast	W.Coast	W.C.Coast	S.W.Coast	S.A.Coast	I.O.Coast	E.C.Coast	N.E.Coast	Madagascar	Islands		

Chapter 3

Soil and terrain suitability for surface irrigation

The evaluation of soil qualities and terrain conditions to predict the performance for specific crops is an essential part of a land evaluation and land use planning exercise applied to agriculture. In the framework of the present study, emphasis was placed on the suitability of soils and terrains for irrigation development.

GENERAL METHODOLOGY

In order to compare soil and terrain conditions with specific crop requirements for optimum growth and production, soil and terrain qualities or characteristics, as derived from the FAO-UNESCO soil map of the world [1], were matched against specific crop requirements derived from agricultural experiments and literature review.

Given the small scale of the soil map of the world (1 : 5 000 000), the approach to evaluating soil and topography effects is the one used in the continental agro-ecological zones study [16]. This method expresses in a qualitative way, using three suitability classes, the estimated performance and suitability of a given land use type, for specific soil and terrain conditions, assuming climatic conditions to be optimum. It proceeds in four steps:

1. Matching crop requirements with the inherent fertility and physical characteristics of each soil unit in the FAO-UNESCO legend of the soil map of the world.

2. Downgrading, if necessary, the soil unit by a factor which takes account of its texture.

3. Downgrading, if necessary, the suitability class obtained after steps 1 and 2 for terrain influences, such as slopes.

4. Downgrading, if necessary, for soil phases such as effective soil depth, the presence of gravel and stones, the presence of high levels of sodium, depending on the specific crop requirements in this respect.

Soil requirements for irrigation

Qualitative land evaluation for irrigation is generally based on interpretation of environmental characteristics, of which slope, soil and groundwater are the most important.

The evaluation criteria adopted here consider surface irrigation using water of good quality (Table 3). Accordingly, some soils considered not suitable for such a development could be

TABLE 3
Criteria used in the evaluation of soil and terrain suitability for irrigation

CRITERIA	CONDITION	UPLAND CROPS	FLOODED RICE
Topography: slope	Optimum	< 2% 2 - 8 %	< 2 % 2 - 8 %
Drainage (1)	Optimum Marginal/Range	W MW - I	P VP - W
Texture (2)	Optimum Range	L - SiCL SL - MCs	CL - MCm SL - MCm
Soil depth	Optimum Marginal	> 100 cm 50 - 100 cm	> 50 cm 20 - 50 cm
Surface stoniness		no stones are acceptable	no stones are acceptable
Subsurface stoniness	Optimum Marginal	< 40 % 40 - 75 %	< 40 % 40 - 75 %
Calcium carbonate	Optimum Marginal	< 30 % 30 - 60 %	< 15 % 15 30 %
Gypsum	Optimum Marginal	< 10 % 10 - 25 %	< 3 % 3 - 15 %
Salinity (3)	Optimum Marginal	< 8 mmhos/cm 8 - 16 mmhos/cm	< 2 mmhos/cm 2 - 4 mmhos/cm
Alkalinity (3,4)	Optimum Marginal	< 15 ESP 15 - 30 ESP	< 20 ESP 20 - 40 ESP

(1) Drainage: W = Well drained; MW = Moderately Well drained; I = Imperfectly drained; P = Poorly drained; VP = Very Poorly drained.

(2) Texture: L = Loamy; SiCL = Silty Clay Loam; SL = Sandy Loam; CL = Clay Loam

(3) Salinity and alkalinity: The criteria refer to salinity and alkalinity conditions that can be accepted for irrigation and possibly improved by irrigation management. The choice of crops has to be made with regard to the local salinity and alkalinity situation.

(4) Alkalinity: ESP = Exchangeable Sodium Percentage.

suitable for sprinkler irrigation or micro-irrigation. It was also assumed that the irrigation infrastructure is in place and that an adequate level of inputs is applied.

Two main land utilization types have been considered: the whole group of upland crops and rice under irrigation. It was decided that, where the soil is suitable for both upland crops and rice, priority be given to rice. This choice is necessary to avoid counting the same land in both categories, thereby artificially increasing the area suitable for irrigation. This approach differs from the 1987 study [20] where the same land was accounted for in both categories.

The attributes of the FAO-UNESCO Soil Map of the World which were used for irrigation appraisal are: topography, drainage, texture, surface and subsurface stoniness, depth, calcium carbonate level, gypsum status, salinity and alkalinity conditions. Criteria were established for evaluating each of these characteristics in relation to the specific requirements for upland crops and flooded rice.

Evaluation techniques

The evaluation of the soil units for irrigation is performed in the same way as the evaluation for rainfed crops [16], namely:

- evaluation of soil units;
- texture modifications;
- slope modifications;
- phase modifications.

RESULTS

Figure 3 shows the detailed map of soil and terrain suitability for surface irrigation per type of crop as a result of the above methodology. Tables 4 and 5 give the extent of suitable land for both types of crops per major basin group and per country respectively. Figure 4 shows the total area of land (rice plus upland crops) suitable for surface irrigation as a percentage of the total area of each of the 24 major basin groups, Figure 5 as a percentage of the area of each of the 53 African countries. As the figures show, the land in the northern part and the southern desert zones is less suitable for surface irrigation than in the rest of Africa.

TABLE 4

Soil and terrain suitability for surface irrigation by major basin group

Basin No. (0)	Area in ha Major Basin group (1)	Total area of the basin group (2)	Soil suitable for irrigation of rice (4)*	Soil suitable for irrigation of upland crops (5)*	Total area of soils suitable for surface irrigation (6)	As % of total area of basin 100*(6)/(2) (7)
01	SENEGAL RIVER	48 318 100	3 491 200	154 600	3 645 800	8
02	NIGER RIVER	227 394 600	28 430 300	513 100	28 943 400	13
03	LAKE CHAD	238 163 500	31 937 700	4 586 900	36 524 600	15
04	NILE	311 236 900	88 784 000	3 235 000	92 019 000	30
05	RIFT VALLEY	63 759 300	11 958 800	1 987 900	13 946 700	22
06	SHEBELLI-JUBA	81 042 700	12 875 700	12 972 200	25 847 900	32
07	CONGO/ZAIRE RIVER	378 905 300	108 785 600	1 029 900	109 815 500	29
08	ZAMBEZI	135 136 500	37 345 600	286 900	37 632 500	28
09	OKAVANGO	32 319 200	6 476 500	135 600	6 612 100	20
10	LIMPOPO	40 186 400	8 987 100	749 000	9 736 100	24
11	ORANGE	89 636 800	9 783 200	408 300	10 191 500	11
12	SOUTH INTERIOR	64 582 600	18 630 900	739 900	19 370 800	30
13	NORTH INTERIOR	580 446 300	18 242 300	30 083 400	48 325 700	8
14	MEDITERRANEAN COAST	67 952 500	7 785 800	4 111 900	11 897 700	18
15	NORTH WEST COAST	67 062 100	5 375 500	7 189 700	12 565 200	19
16	WEST COAST	143 019 600	26 329 800	3 237 600	29 567 400	21
17	WEST CENTRAL COAST	70 477 400	16 292 000	36 400	16 328 400	23
18	SOUTH WEST COAST	51 620 000	10 665 200	3 127 300	13 792 500	27
19	SOUTH ATLANTIC COAST	36 548 500	2 201 600	1 840 300	4 041 900	11
20	INDIAN OCEAN COAST	66 378 500	13 569 400	1 282 900	14 852 300	22
21	EAST CENTRAL COAST	102 625 200	23 733 300	1 108 700	24 842 000	24
22	NORTH EAST COAST	72 570 200	6 044 400	5 734 700	11 779 100	16
23	MADAGASCAR	58 704 000	14 138 900	358 500	14 497 400	25
24	ISLANDS	934 600	134 100	50 400	184 500	20
	Total for Africa	3 029 020 800	511 998 900	84 961 100	596 960 000	20

* In order to be able to add up the figures of soil and terrain suitability for rice and upland crops, it was decided that priority be given to rice where the suitability was the same for both crops.

TABLE 5
Soil and terrain suitability for surface irrigation by country

Area in ha Country (1)	Total area of the country (2)	Soil suitable for irrigation of rice (3)*	Soil suitable for irrigation of upland crops (4)*	Total area of soils suitable for surface irrigation (5)	As % of total area of country (5/2)*100 (6)
ALGERIA	238 174 000	8 482 900	19 467 600	27 950 500	12
ANGOLA	124 670 000	22 796 600	4 187 500	26 984 100	22
BENIN	11 262 000	3 738 600	0	3 738 600	33
BOTSWANA	58 173 000	13 189 400	5 400	13 194 800	23
BURKINA FASO	27 400 000	5 438 000	0	5 438 000	20
BURUNDI	2 783 400	302 100	286 700	588 800	21
CAMEROON	47 544 000	11 784 600	26 800	11 811 400	25
CAPE VERDE	403 000	9 900	39 000	48 900	12
CENTRAL AFRICAN REP.	62 298 000	7 704 500	0	7 704 500	12
CHAD	128 400 000	20 589 200	5 077 500	25 666 700	20
COMOROS	186 100	16 900	0	16 900	9
CONGO	34 200 000	9 257 600	45 600	9 303 200	27
COTE D'IVOIRE	32 246 200	4 545 300	1 050 700	5 596 000	17
DJIBOUTI	2 320 000	246 100	50 600	296 700	13
EGYPT	100 145 000	6 477 400	655 900	7 133 300	7
EQUATORIAL GUINEA	2 805 000	919 000	0	919 000	33
ERITREA	12 189 000	1 703 000	2 565 400	4 268 400	35
ETHIOPIA	110 001 000	20 918 100	9 418 300	30 336 400	28
GABON	26 767 000	5 816 000	0	5 816 000	22
GAMBIA	1 130 000	495 100	0	495 100	44
GHANA	23 854 000	5 684 500	5 000	5 689 500	24
GUINEA	24 585 700	3 980 100	473 900	4 454 000	18
GUINEA BISSAU	3 612 000	603 700	0	603 700	17
KENYA	58 037 000	11 405 600	5 979 100	17 384 700	30
LESOTHO	3 035 000	652 000	0	652 000	21
LIBERIA	9 775 000	1 129 200	1 036 200	2 165 400	22
LIBYA	175 954 000	7 915 800	4 914 000	12 829 800	7
MADAGASCAR	58 704 000	14 138 900	358 500	14 497 400	25
MALAWI	11 848 000	2 467 200	0	2 467 200	21
MALI	124 019 000	9 939 600	202 200	10 141 800	8
MAURITANIA	102 552 000	2 462 000	6 325 300	8 787 300	9
MAURITIUS	204 000	29 000	0	29 000	14
MOROCCO+W.SAHARA	71 250 000	6 622 800	7 813 400	14 436 200	20
MOZAMBIQUE	80 159 000	17 432 000	983 700	18 415 700	23
NAMIBIA	82 490 000	11 111 700	2 133 900	13 245 600	16
NIGER	126 700 000	3 476 000	86 100	3 562 100	3
NIGERIA	92 377 000	18 080 700	317 900	18 398 600	20
RWANDA	2 634 000	220 600	80 300	300 900	11
SAO TOME & PRINCIPE	96 000	10 700	0	10 700	11
SENEGAL	19 672 000	2 742 500	290 200	3 032 700	15
SEYCHELLES	45 500				
SIERRA LEONE	7 174 000	985 000	724 500	1 709 500	24
SOMALIA	63 766 000	8 361 500	4 427 600	12 789 100	20
SOUTH AFRICA	122 104 000	21 434 600	1 163 600	22 598 200	19
SUDAN	250 581 000	66 955 100	1 814 100	68 769 200	27
SWAZILAND	1 736 400	339 500	0	339 500	20
TANZANIA	94 509 000	23 344 700	908 700	24 253 400	26
TOGO	5 678 500	1 114 700	0	1 114 700	20
TUNISIA	16 361 000	1 625 500	1 227 900	2 853 400	17
UGANDA	23 588 000	7 652 000	23 700	7 675 700	33
ZAIRE	234 486 000	78 728 100	9 700	78 737 800	34
ZAMBIA	75 261 000	26 540 700	2 400	26 543 100	35
ZIMBABWE	39 076 000	10 382 600	782 200	11 164 800	29
Total for Africa	3 029 020 800	511 998 900	84 961 100	596 960 000	20

* In order to be able to sum the figures of soil and terrain suitability for rice and upland crops, it was decided that priority be given to rice where the suitability was the same for both crops. For this reason the figures given in column 4 must be considered as a lower limit for soil suitability to irrigated upland crops.

FIGURE 3
Soil and terrain suitability for surface irrigation

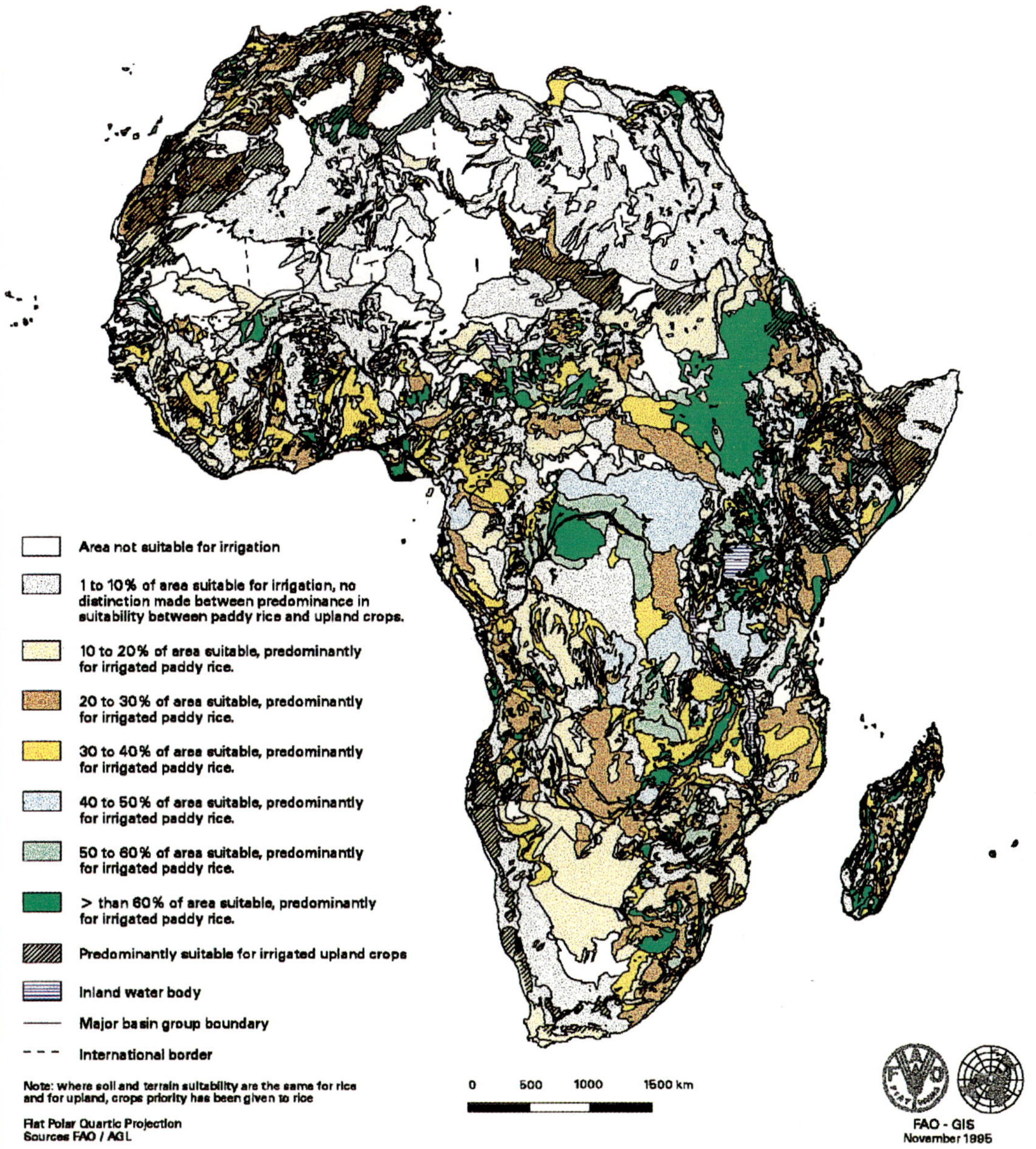

FIGURE 4
Extent of land siutable for surface irrigation (as % of basin area)

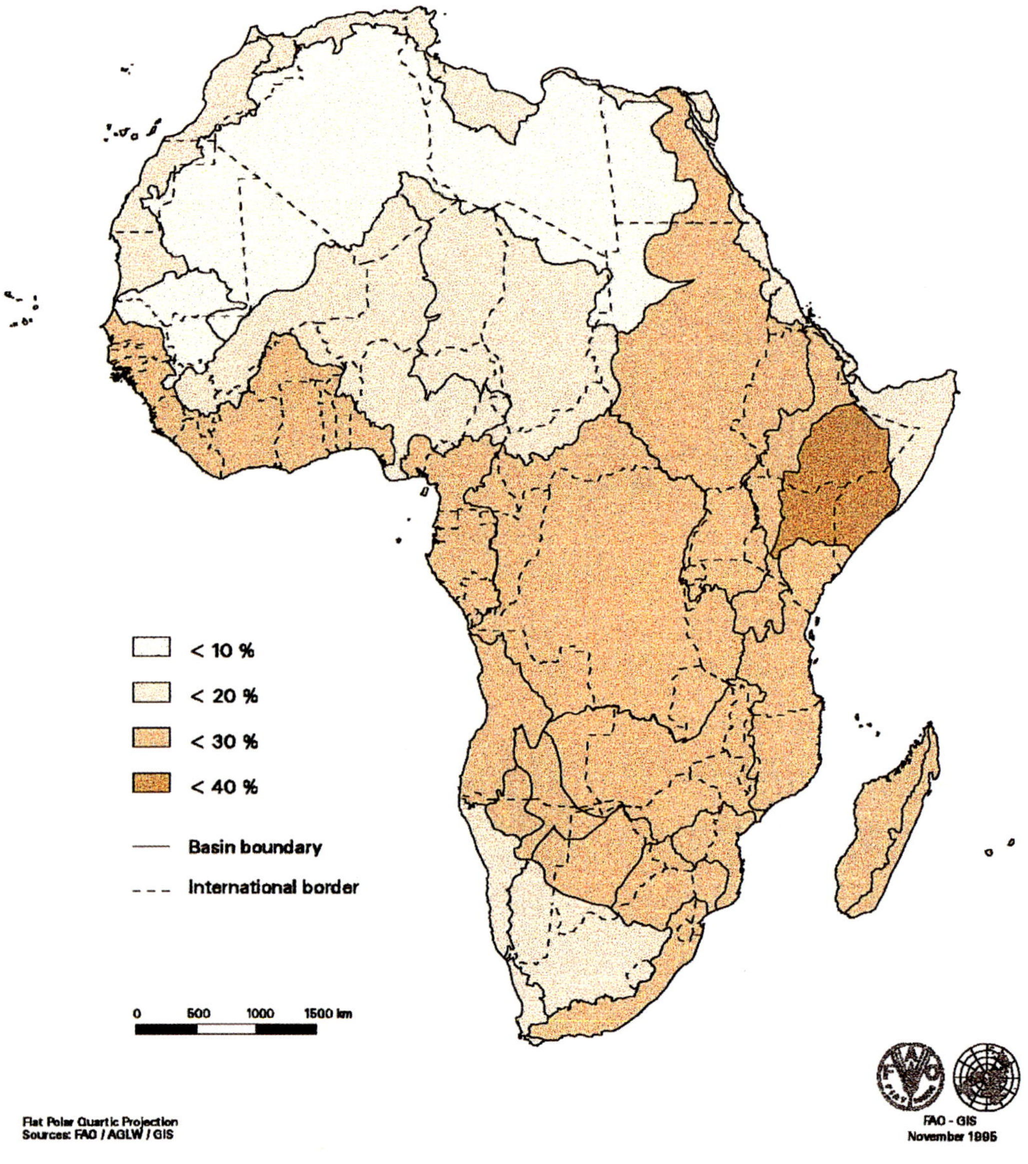

FIGURE 5
Extent of land suitable for surface irrigation (as % of country area)

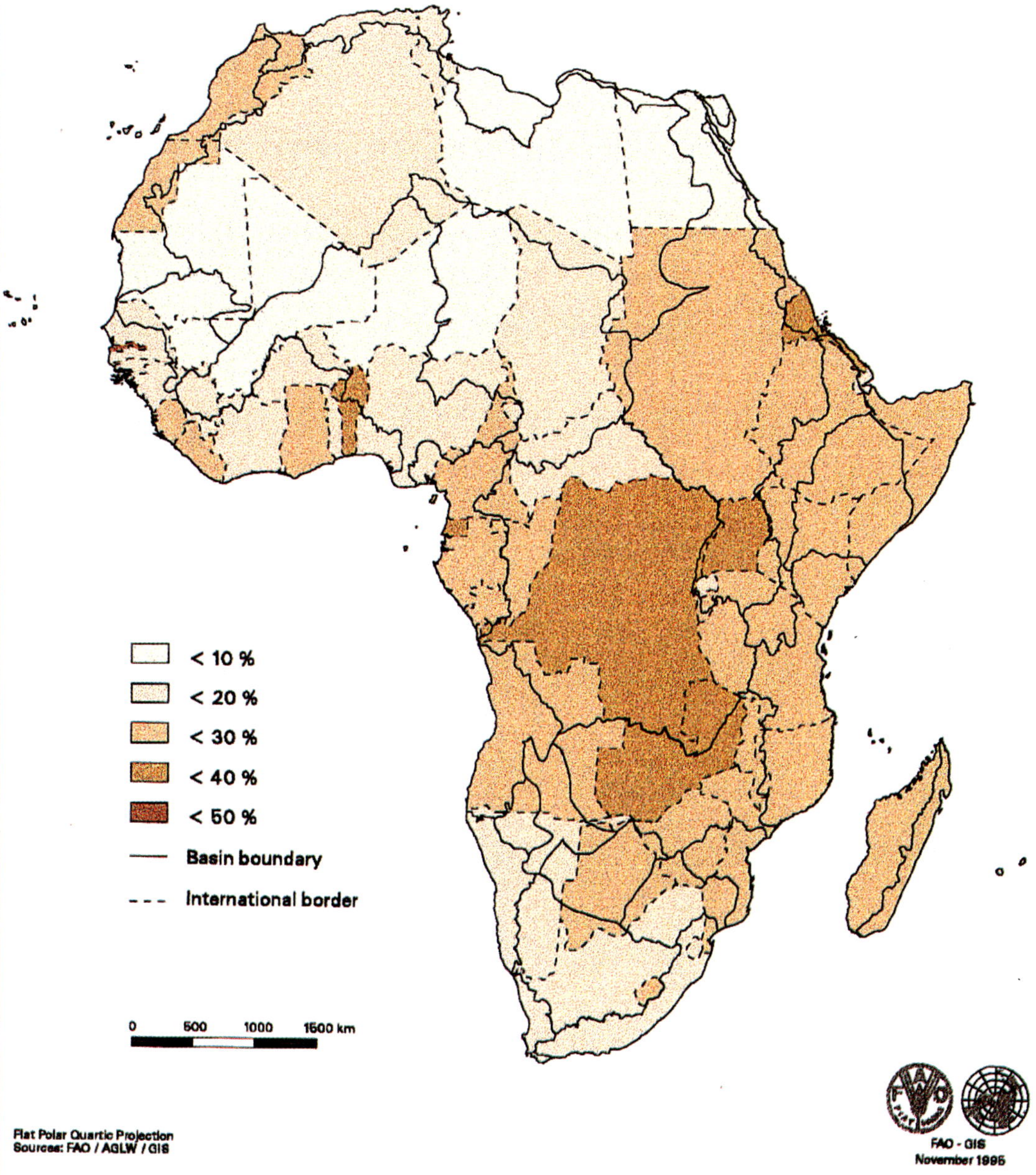

Chapter 4

Water resources

Assessment of water resources can only be done at basin level. At country level it is possible to assess that part of the water resources which is generated inside the borders of the country. However, exchanges of water through international rivers represent a significant part of the water balance for several countries. In extreme cases, an arid country may depend almost entirely on water produced outside its borders. This explains the necessity to compute irrigation potential on the basis of river basins rather than countries.

In 1995 FAO conducted a review of the water resources of the African countries, considering internally as well as globally produced renewable water resources [21]. The survey was principally based on information produced by countries or regional and international organizations, completed with information gathered from previous studies. A summary of the review is given below.

METHODOLOGY AND DEFINITIONS USED

Potential yield

Potential yield is defined here as the global amount of water resources, be it surface water or groundwater, which is generated on a yearly basis in a given area.

Surface water and groundwater

The most widely used approach to computing water resources at national level is to study surface water and groundwater resources separately. One of the major risks in assessing them separately lies in the possible double counting of part of the resources. For example, in countries where part of the river flow is generated by discharges from the upper aquifers, mostly in humid areas, this figure includes a part of the water resources which can be considered as groundwater and could in fact be developed through wells. On the other hand, in arid areas, the river system usually acts as a preferential source for groundwater recharge and shows very limited base flow. The river runoff typically occurs in flash floods of high intensity and short duration. In this survey special care was taken with the computation methods, so that possible overlaps could be detected and removed from the accounting of water resources.

Internally and globally produced renewable water resources

When computing water resources on a country basis, it is important to make a distinction between internally and globally produced renewable water resources[1]. Internally produced renewable water resources (IRWR) refer to the water resources resulting from rain falling within the borders of the country and are a combination of surface water and groundwater resources. Globally produced renewable water resources (GRWR) are obtained by adding incoming surface water and groundwater flows to the internally produced renewable water resources.

The internal water resources figures are the only quantities that can be added together for regional or continental assessment. The computation of global renewable water resources requires the assessment of surface water and groundwater flowing from neighbouring countries and between neighbouring countries (rivers that form the border between countries). Rules have been established for these computations, which are explained in detail in [21]. By definition, global water resources are not additive at the scale of international river basins. The definition implies that unused water, accounted for as a resource in upstream countries, is again counted as a resource in downstream countries.

Periods of reference

The review concentrated on long-term averages and did not consider seasonal or inter-annual variations. However, it should be stressed that the review is based on information available from a multitude of sources and that no consistency in the choice of the period of reference can be expected. Some examples of rather important differences that might occur in average flow estimates, depending on the period of reference, are given in Chapter 6 for specific basins. Chapter 6 also gives more details on each of the 25 major basin groups, including information on their rivers, discharges and seasonal flow variations.

Evaporation from wetlands and lakes

In humid regions, the internally generated water resources of a country can be calculated by comparing incoming and outgoing flows and taking into account withdrawals inside the country. In arid regions, however, this method leads to important underestimates and even negative values for internally produced water resources. This situation occurs, for example, in Sudan, Mali and Botswana, where the quantity of water leaving the country is inferior to the quantity of water flowing into the country. In such countries losses by evaporation play a major role and a country-wide approach is not feasible. In arid regions, groundwater recharge also plays a major role in the assessment of water resources. In such situations, it is necessary to review in detail all possible sources of water and assess the quantity of water which would be available before being lost by evaporation.

In all climate situations it is difficult to account for evaporation from large lakes in the water balance of a country. This uncertainty may greatly impair the reliability of a country estimate. This typically the case of Lake Victoria for which no consistent water balance can be established.

---

1 The term 'renewable' here is used as opposed to fossil waters, which have a negligible rate of recharge on the human scale and can thus be considered 'non-renewable'. Non-renewable resources are usually expressed either in terms of volumes or extractable flow, while renewable resources are always a measure of flow, usually presented on a yearly basis.

In the present review, no systematic approach could be taken towards evaporation from lakes or other water bodies. Sometimes the resources were calculated without removing evaporation losses, as was the case for Mali, Uganda and Egypt[1]. For Sudan, evaporation in wetlands was subtracted from the total to obtain internal water resources.

RESULTS

Table 6 and Figure 6 present the results of the review in terms of water resources by country. Surface water and groundwater resources have been presented in a non-additive way, that is to say that the base flow appears in both columns. This 'overlap' represents the part of water resources which is common to surface water and groundwater. The reason for presenting these figures in such a way is that this is how the water resources are usually presented in country studies and that there is no objective reason for subtracting the common part from one or other category.

The total for internally produced renewable water resources in column 4 is found by adding the surface water and groundwater resources and then subtracting the overlap (base flow) to avoid double counting.

Global renewable water resources are the sum of internal renewable water resources and incoming water. In an attempt to make a distinction between flow entering a country and border rivers, these two components have been presented in two separate columns (5 and 6).

In order to complete the picture on water resources, non-conventional sources of water, including potential development of fossil resources and desalination, have been added to the table. This may be of particular relevance to arid regions. However, it must be stressed that the figures in these columns are indicative and should be subjected to more detailed study. Moreover, technological advances can also lead to different estimates of the potential use of desalinated or fossil water.

BREAKDOWN OF WATER RESOURCES BY BASIC UNIT

The above review of water resources, based on countries, does not provide information about the distribution of the resources among the various river basins and basic units. For several countries, where detailed studies have been carried out, this information exists and was used in the assessment of water resources and irrigation potential. For several other countries, however, the information was not available.

A systematic approach, based on information available through FAO's Geographic Information System (GIS) has thus been used to provide information on water resources for the units for which it was not available in the literature.

A first estimate of water resources by basic unit can be obtained by multiplying annual precipitation P by a runoff coefficient c.

---

1 The external incoming water resources of Egypt are estimated at 65.5 km^3/year. However, the evaporation from the Aswan reservoir, just downstream of the border with Sudan, is estimated at 10 km^3/year, so the flow at the outlet of the reservoir is in fact only 55.5 km^3/year.

$$Q = c.P$$

where Q is the average annual flow produced inside the basic unit; Q and P are expressed in mm/year and c is dimensionless.

This was achieved by preparing a raster coverage of runoff coefficients from a map of African runoff coefficients [28] and combining the results with the annual average rainfall map (Figure 7) [23].

By multiplying the precipitation map by the runoff coefficient map, a map of runoff Q was obtained. In first approximation, this runoff can be assimilated to internal renewable water resources. This approximation is specially valid in humid areas. In arid areas, where groundwater resources are relatively important compared to surface water, this approximation may be less valid.

Integration of the runoff figures at the level of each country was performed to obtain country values of runoff, R, expressed in km^3/year. The 53 country values of R were then compared with the figures of IRWR in Table 6. To avoid giving excessive importance to large and humid countries, the results were plotted in a logarithmic scale after having transformed IRWR and R from km^3/year to mm/year by dividing it by the area of the country. The comparison showed good agreement between the two sets of data and it was decided that this method was satisfactory to provide estimates of internal water resources for those basic units for which no information was available.

A similar test was performed to compare runoff measured at the outlet of large basins (obtained from the literature) and the value of runoff computed by integrating Q over the basins. The results show a systematically lower value of the measured runoff R_m compared to R, the relative difference being more important in arid than in humid areas. This can easily be explained by the losses occurring in the basins before the water reaches the outlet: losses by evaporation in the reaches, lakes, wetland and uses by agriculture and other sectors. Losses are relatively more important in arid areas, where the evaporation potential is higher and use by agriculture is usually more important than in humid areas. However, relative uncertainties with arid regions are less important because most of the missing information concerns humid countries.

The results of the second test (basin level) also bring the concept of scale in assessment of water resources. As both the runoff coefficient map [28] and the IRWR figures in Table 6 [21] used the country as a basis for assessment of water resources, they give similar results. A study at a smaller scale (local level) would probably show higher values of measured runoff due to reduced possibilities of losses, against the lower values measured at the larger scale (basin level).

TABLE 6
Water resources by country (all figures in km^3/yr)

	COUNTRY	Internal renewable water resources					Incoming water			Global renewable water res.			Other resources	
		Surface water	Groundwater		Overlap	Total	Surface water		Groundwater	Surface water	Groundwater	Total	Desalination	Groundw. depletion
				C.		(1+2-3)	T	B		(1+5+6)	(2+7)	(8+9-3)		
		(1)	(2)		(3)	(4)	(5)	(6)	(7)	(8)	(9)	(10)	(11)	(12)
1	ALGERIA	13.2	1.7	a	1	13.9	0.4	0	0.03	13.6	1.73	14.33	0.03	5
2	ANGOLA	182	72	b	70	184	0	0	0	182	72	184	x	0
3	BENIN	10	1.8	a	1.5	10.3	0.5	15	0	25.5	1.8	25.8	0	0
4	BOTSWANA	1.7	1.7	a	0.5	2.9	11.5	0.3	0	13.5	1.7	14.7	0	0
5	BURKINA FASO	13	9.5	a	5	17.5	...	x	0	x+13	9.5	x+17.5	0	0
6	BURUNDI	3.5	2.1	b	2	3.6	x	0	0	x+3.5	2.1	x+3.6	0	0
7	CAMEROON	268	100	b	100	268	...	0	0	268	100	268	0	0
8	CAPE VERDE	0.18	0.12	a	0	0.3	0	0	0	0.18	0.12	0.3	x	0
9	CENTRAL AFRICAN REP.	141	56	b	56	141	0	x	0	x+141	56	x+141	0	0
10	CHAD	13.5	11.5	b	10	15	28	0	0	41.5	11.5	43	0	0
11	COMOROS	x	x		x	1.02	0	0	0	x	x	1.02	0	0
12	CONGO	222	198	b	198	222	...	610	0	832	198	832	...	0
13	COTE D'IVOIRE	74	37.7	a	35	76.7	1	x	0	x+75	37.7	x+77.7	0	0
14	DJIBOUTI	x	x		x	0.3	0	2	x	x+2	x	x+2.3	0	0
15	EGYPT	0.5	1.3	a	0	1.8	65.5	0	1.2	66	2.5	68.5	0.01	...
16	EQUATORIAL GUINEA	25	10	b	5	30	...	0	0	25	10	30	0	0
17	ERITREA	x	x		x	2.8	0	6	0	x	x	8.8	0	0
18	ETHIOPIA	x	x		x	110	0	0	0	x	x	110	0	0
19	GABON	162	62	b	60	164	0	0	0	162	62	164	0	0
20	GAMBIA	3	0.5	b	0.5	3	5	0	0	8	0.5	8	0	0
21	GHANA	29	26.3	a	25	30.3	22.9	0	0	51.9	26.3	53.2	0	0
22	GUINEA	226	38	b	38	226	0	0	0	226	38	226	0	0
23	GUINEA-BISSAU	12	14	b	10	16	11	0	0	23	14	27	0	0
24	KENYA	17.2	3	a	0	20.2	...	10	0	27.2	3	30.2	0	0
25	LESOTHO	4.73	0.5	b	0	5.23	0	0	0	4.73	0.5	5.23	0	0
26	LIBERIA	200	60	b	60	200	32	0	0	232	60	232	0	0
27	LIBYA	0.1	0.5	a	0	0.6	0	0	0	0.1	0.5	0.6	0.003	2 to 4
28	MADAGASCAR	332	55	b	50	337	0	0	0	332	55	337	0	0
29	MALAWI	16.14	1.4	b	0	17.54	1.14	0	0	17.28	1.4	18.68	0	0
30	MALI	50	20	a	10	60	40	0	0	90	20	100	0	0
31	MAURITANIA	0.1	0.3	a	0	0.4	0	11	0	11.1	0.3	11.4	x	0
32	MAURITIUS	2.03	0.68	a	0.5	2.21	0	0	0	2.03	0.68	2.21	0	0
33	MOROCCO	22.5	7.5	a	0	30	0	0	0	22.5	7.5	30	0.004	0
34	MOZAMBIQUE	97	17	b	17	97	106	5	0	208	17	208	0	0
35	NAMIBIA	4.1	2.1	b	0	6.2	11.3	28	0	43.4	2.1	45.5	0.003	0
36	NIGER	1	2.5	a	0	3.5	29	0	0	30	2.5	32.5	0	0
37	NIGERIA	214	87	b	80	221	59	x	0	x+273	87	x+280	0	0
38	RWANDA	5.2	3.6	b	2.5	6.3	0	x	0	x+5.2	3.6	x+6.3	0	0
39	SAO TOME AND PRINCIPE	x	x		x	2.18	0	0	0	x	x	2.18	0	0
40	SENEGAL	23.8	7.6	b	5	26.4	2	11	0	36.8	7.6	39.4	0	0
41	SEYCHELLES	x	x		x	x	0	0	0	x	x	x	0	0
42	SIERRA LEONE	150	50	b	40	160	0	0	0	150	50	160	0	0
43	SOMALIA	5.7	3.3	b	3	6	7.5	0	x	13.2	x+3.3	x+13.5	0	0
44	SOUTH AFRICA	40	4.8		0	44.8	5.2	0	0	45.2	4.8	50	x	0
45	SUDAN	28	7		0	35	119	0	0	147	7	154	0	0
46	SWAZILAND	x	x		x	2.64	1.87	0	0	x+1.87	x	x+4.51	0	0
47	TANZANIA	80	30	b	30	80	0	9	0	89	30	89	0	0
48	TOGO	10.8	5.7	a	5	11.5	0.5	0	0	11.3	5.7	12	0	0
49	TUNISIA	2.31	1.21	a	0	3.52	0.32	0	0.1	2.63	1.31	3.94	0.009	1
50	UGANDA	35	29	b	25	39	27	0	0	62	29	66	0	0
51	ZAIRE	934	421	b	420	935	84	x	0	x+1018	421	x+1019	0	0
52	ZAMBIA	33.1	47.1		0	80.2	35.8	x	0	x+68.9	47.1	x+116	0	0
53	ZIMBABWE	13.1	5	b	4	14.1	0	x+5.9	0	x+19	5	x+20	0	0
	TOTAL	3721	1517		1370	3988								

C: Method of computing groundwater: a = recharge of the aquifers, b = baseflow of river system;
T: transboundary flow; B: bordering river; x: unknown; ...: negligible.

FIGURE 6
Water resources by country (all figures in km³/year)

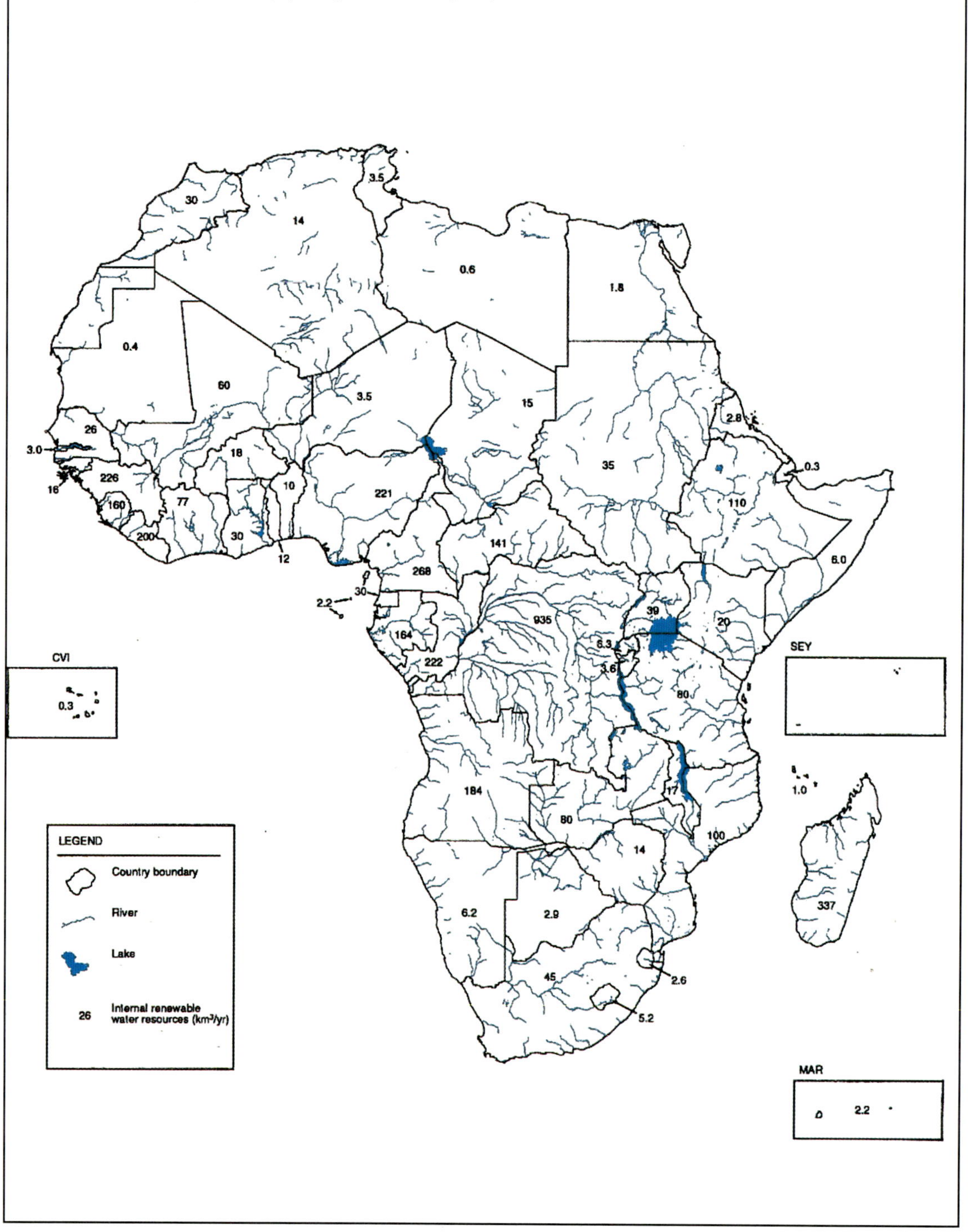

FIGURE 7
Average annual rainfall

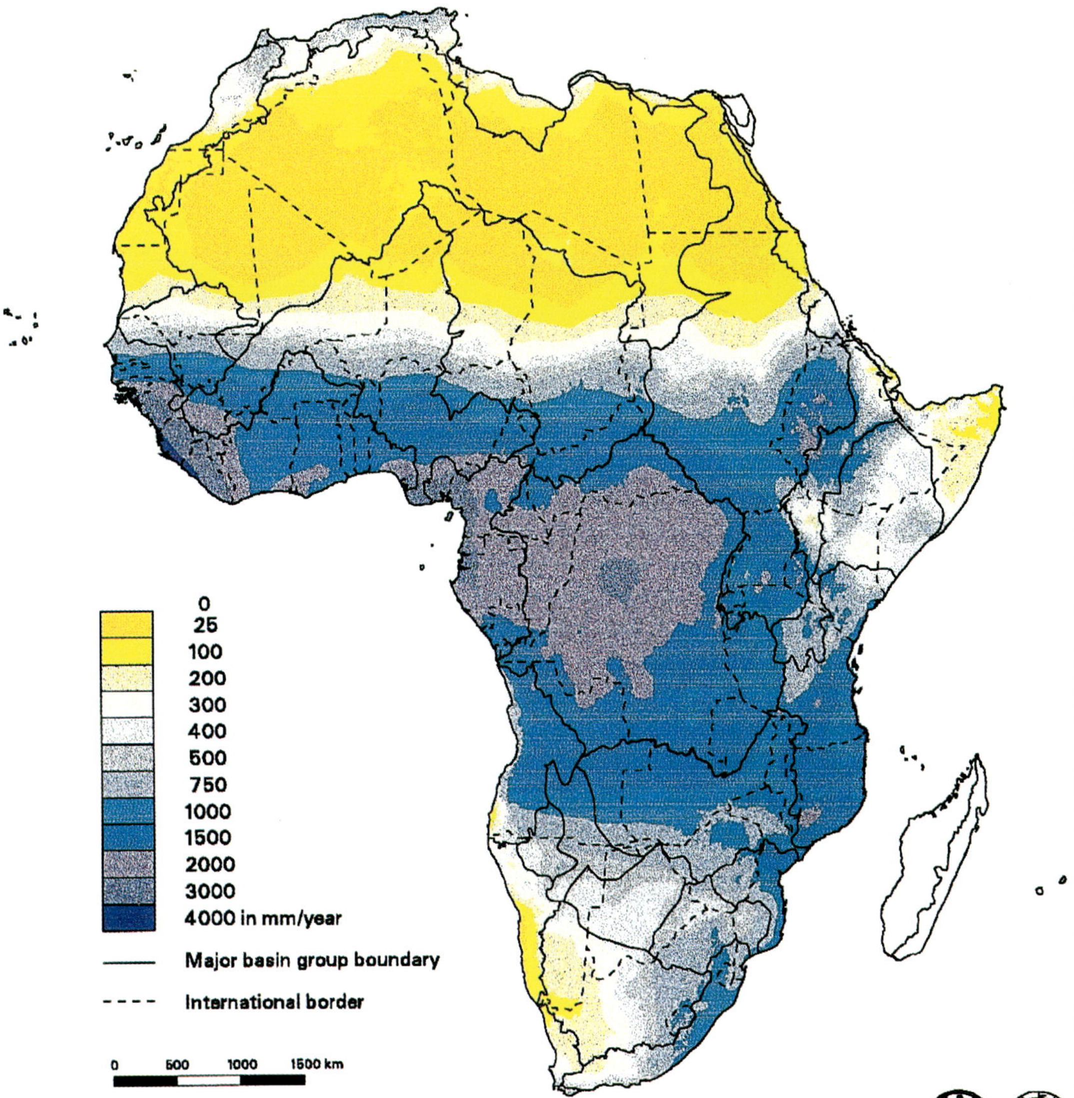

Flat Polar Quartic Projection
Sources:FAO /M.Hutchinson,National Australian University

FAO - GIS
November 1995

Chapter 5

Irrigation water requirements

The assessment of the irrigation potential, based on soil and water resources, can only be done by simultaneously assessing the irrigation water requirements (IWR) (Figure 1).

Net irrigation water requirement (NIWR) is the quantity of water necessary for crop growth. It is expressed in millimetres per year or in m^3/ha per year (1 mm = 10 m^3/ha). It depends on the cropping pattern and the climate. Information on irrigation efficiency is necessary to be able to transform NIWR into gross irrigation water requirement (GIWR), which is the quantity of water to be applied in reality, taking into account water losses. Multiplying GIWR by the area that is suitable for irrigation gives the total water requirement for that area. In this study water requirements are expressed in km^3/year.

Calculations of irrigation water requirements are done while preparing national water master plans or irrigation projects. Useful information was obtained from a number of country studies available from AQUASTAT [21a], but the information was based on many different approaches. For the purpose of this study the need was felt to develop a method of computing irrigation water requirements for the whole continent in a systematic way. In order to be able to do this at the scale of the continent, assumptions have to be made on the definition of areas to be considered homogeneous in terms of rainfall, potential evapotranspiration, cropping pattern, cropping intensity and irrigation efficiency.

METHODOLOGY

For the calculation of irrigation water requirements the following steps have been followed:

- Delineation of major irrigation cropping pattern zones. These zones are considered homogeneous in terms of types of irrigated crops grown, crop calendar, cropping intensity and gross irrigation efficiency. Represented on the map of Africa, they should be viewed as regions where some homogeneity can be found in terms of irrigated crops. The cropping pattern proposed for the zone should be viewed as representative of an 'average' rather than a 'typical' irrigation scheme.

- Definition of the area of influence of the climate stations (in GIS) and quality check on the climate data.

- Combination of the irrigation cropping pattern zones with the climate stations' zones (in GIS) to obtain basic mapping units.

- Calculation of net and gross irrigation water requirements for different scenarios.

- Comparison with existing data and final adjustment.

Delineation of irrigation cropping pattern zones

The criteria used for the delineation of the irrigation cropping pattern zones were, in order of decreasing importance: distribution of irrigated crops, average rainfall trends and patterns, topographic gradients, presence of large river valleys (Nile, Niger, Senegal), presence of extensive wetlands (the Sudd in Sudan), population pressure, technological differences and crop calendar above and below the equator (Zaire).

The starting point was the type of irrigated crops currently grown in Africa. This resulted in 18 zones. From these zones, sub-zones showing a different cropping intensity or a different crop calendar were defined. This resulted in a total of 24 irrigation pattern zones (Figure 8), which are considered to be homogeneous for:

- crops currently grown;
- crop calendar;
- cropping intensity.

Only the main crops currently grown, those occupying at least 85% of the irrigated area, were considered. Land occupation of the remaining 15% by secondary crops was assigned to the main crops.

An 'average' typical monthly crop calendar was assigned to each zone, based on work done by FAO's global information and early warning system, and on information from the reference library of FAO's agrometeorology group, AQUASTAT and, for eastern Africa, from the IGADD crop production system zones inventory.

For each crop the actual cropping intensity was derived from national crop production and land use figures extracted from the FAO AGROSTAT [6] and AQUASTAT [21a] databases. It ranges from 100 to 200%, according to the crop calendar. The cropping intensity to be used in this study of irrigation potential ('potential' scenario) was generally estimated by increasing current values by 10 to 20%, but it was assumed that because of market limitations the current high intensity (in relative terms) of vegetables in certain parts of the continent would not be found in the potential scenario. Therefore, intensities of cereal crops are higher in the potential scenario than in the actual situation.

Table 7 summarizes the cropping pattern, crop calendar and cropping intensities for the 24 zones used in this study.

Definition of the climate stations' area of influence

The climate data from the FAOCLIM cd-rom were used, as this was the most up to date climate database available [7]. This data set includes long term average rainfall and reference potential evapotranspiration (ET_0) data for 1025 stations throughout Africa. ET_0 was calculated by the Penman-Monteith method [4].

To obtain a spatial coverage of climate data (P, ET_0) over the continent, each station was assigned an area of influence using the Thiessen polygons method. This method assigns an area of 'nearest vicinity' to each climate station. Figure 9 gives an indication of the density of the stations over the continent. As expected, the desert areas in northern and southern Africa are much less well covered than the rest of the continent. The rainfall data were compared with raster maps prepared by the Australian National University [23] and corrected where necessary.

Combination of cropping pattern zones with the climate stations

In ArcInfo, the 24 cropping pattern zones and the 1025 climate station data were merged. This resulted in 1437 basic map features, homogeneous in irrigation cropping characteristics and climate. All further calculations were carried out on these 1437 basic mapping units.

Calculation of irrigation water requirements

Crop water requirements (CWR) for a given crop, i, are given by:

$$CWR_i = \sum_{t=0}^{T} (kc_{i_t} \cdot ET_{o_t} - P_{eff_t}) \qquad \text{unit: mm}$$

where kc_{it} is the crop coefficient of the given crop i during the growth stage t and where T is the final growth stage.

Each crop has its own water requirements. Net irrigation water requirements (NIWR) in a specific scheme for a given year are thus the sum of individual crop water requirements (CWR_i) calculated for each irrigated crop i. Multiple cropping (several cropping periods per year) is thus automatically taken into account by separately computing crop water requirements for each cropping period. By dividing by the area of the scheme (S, in ha), a value for irrigation water requirements is obtained and can be expressed in mm or in m^3/ha (1 mm = 10 m^3/ha).

$$NIWR = \frac{\sum_{i=1}^{n} CWR_i \cdot S_i}{S} \qquad \text{unit: mm}$$

where S_i is the area cultivated with the crop i in ha.

The cropping intensity of the scheme can be defined as:

$$\frac{\sum_{i=1}^{n} S_i}{S}$$

FAO's CROPWAT software (version 5.7) was used to compute NIWR for each of the 137 basic units described in chapter 2 [3]. The model was run for three different scenarios:

- actual cropping intensity, effective rainfall[1];
- potential cropping intensity, effective rainfall;
- potential cropping intensity, dependable rainfall[2].

---

1 Effective rainfall was computed according to the "USDA Soil Conservation Service Method" formula in [3], page 21.

2 Dependable rainfall, the combined effect of dependable rainfall (80% probability of exceedance) and estimated 'losses' due to runoff and percolation, was calculated according to the formula in [3], page 21.

Gross irrigation water requirement (GIWR) is the amount of water to be extracted (by diversion, pumping) and applied to the irrigation scheme. It includes NIWR plus water losses:

$$GIWR = \frac{1}{E} . NIWR \qquad unit: mm$$

where E is the global efficiency of the irrigation system.

Limited objective information on irrigation efficiency was available and estimates were based on several criteria:

- figures found in literature;
- type of crops irrigated;
- the level of intensification of the irrigation techniques.

In this study the irrigation efficiencies for the 'potential' scenario range from 45 to 80% (Table 8).

Point observations were further generalized by hand to obtain zones of homogeneous irrigation water requirements (HIWR). A total of 84 HIWR zones were defined.

The methodology was tested and calibrated using a case study on the Egyptian Nile Valley and Delta where water requirements and availability could be computed with relative precision.

RESULTS

Figures 10 and 11 show the net and gross irrigation water requirements for potential cropping pattern and potential irrigation efficiency with effective rainfall. Table 8 summarizes the figures for each of the 84 zones. NIWR and GIWR for the potential scenario with effective rainfall were further combined with the 136 basic units of this study to obtain individual NIWR and GIWR for each of these units through GIS.

The results have been compared with figures available from country studies (national water master plans, projects, etc.). The comparison shows that the methodology yields relatively accurate regional estimates of IWR that are suitable for the present study. Discrepancies with country studies find their origin mostly in the assumptions made on cropping pattern, cropping intensity and irrigation efficiency, and are discussed in details in Chapter 6.

The influence of *cropping pattern zones* on the quality of the output is of prime importance. Important differences in irrigation water requirements in adjacent zones is one of the consequences of this approach. For instance, in Burkina Faso, areas located north of the 1000-mm annual rainfall line have a gross potential water requirement of 500 mm per year, while areas located just south of this line need more than 2 800 mm per year. This artificial break is due to the choice of the irrigated cropping pattern zones, where it was decided that no rice was cultivated under 1 000 mm of rainfall per year. Within the cropping pattern zones, the boundaries of irrigation water requirement zones follow rainfall trends.

The estimates used for *cropping intensity* and *irrigation efficiencies*, to obtain the gross irrigation water requirements from the net figures, also have a direct influence on the results presented on the final maps.

The differences in irrigation water requirements between adjacent zones is directly related to the *density of the climate stations' network*. In low-density areas, such as the Sahara and southern Africa, differences in IWR between adjacent zones are high (up to 600 mm/year gross requirements) as the low station density does not allow the delineation of HIWR zones with smaller differences. A high density of the station network in the rest of the continent, in combination with rainfall raster maps, has resulted in differences of a maximum of 200 mm/year gross requirements between adjacent zones.

FIGURE 8
Irrigation cropping pattern zones. List of cropping pattern zones.

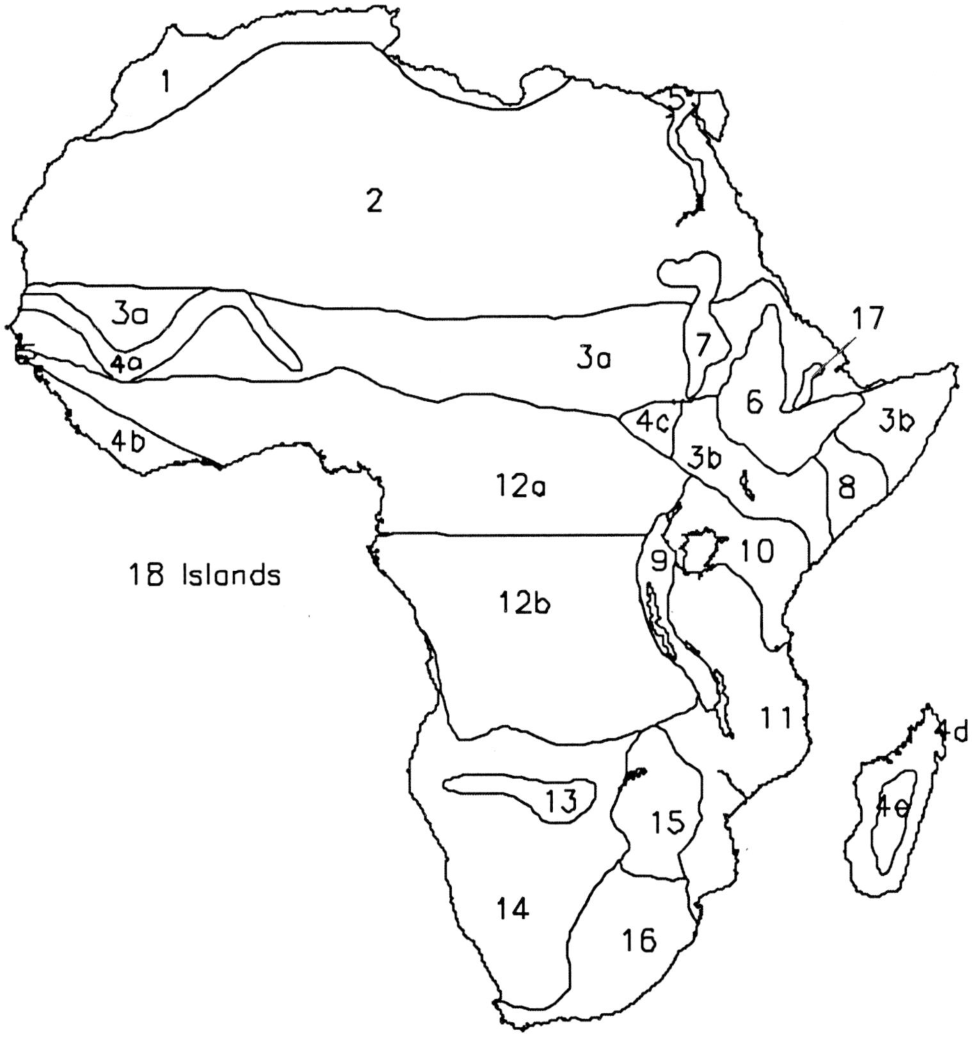

1	Mediterranean coastal zone
2	Saharan oases
3a	Semi-arid to arid savannas in West-East Africa
3b	Semi-arid/arid savanna (Somalia - Kenya - Southern Sudan)
4a	Rice : Niger/Senegal rivers
4b	Rice : Gulf of Guinea
4c	Rice : Southern Sudan
4d	Rice : Madagascar tropical lowland
4e	Rice : Madagascar highland
5	Egyptian Nile and delta
6	Ethiopian highlands
7	Sudanese Nile area
8	Shebelli-Juba river area in Somalia
9	Rwanda - Burundi - Southern Uganda highlands
10	Southern Kenya - Northern Tanzania
11	Malawi - Mozambique - Southern Tanzania
12a	West and Central African humid areas above the equator
12b	Central African humid areas below the equator
13	River affluents on Angola - Namibia - Botswana border
14	South Africa - Namibia - Botswana desert and steppe
15	Zimbabwe highland
16	South Africa - Lesotho - Swaziland
17	Awash river area in Ethiopia
18	All islands (Comoros - Mauritius - Seychelles - Cape Verde)

TABLE 7
Irrigation cropping patterns for the 24 zones

1. Mediterranean coastal zone

cropping season	main crops	J	F	M	A	M	J	J	A	S	O	N	D	cropping intensity actual	cropping intensity potential
summer	vegetables		p	-	-	-	-	-	-	-	-	h		40	40
winter	wheat	-	-	-	-	h						p	-	15	15
winter	fodder	-	-	-	h						p	-	-	25	25
all year	arboriculture	-	-	-	-	-	-	-	-	-	-	-	-	20	20
														100	100

2. Saharan oases

cropping season	main crops	J	F	M	A	M	J	J	A	S	O	N	D	cropping intensity actual	cropping intensity potential
summer	vegetables			p	-	-	-	-	-	-	-	h		30	30
winter	wheat	-	-	-	-	h						p	-	30	30
winter	fodder	-	-	-	-	h						p	-	20	20
all year	arboriculture	-	-	-	-	-	-	-	-	-	-	-	-	20	20
														100	100

3a. Semi-arid to arid savannas in West-East Africa

cropping season	main crops	J	F	M	A	M	J	J	A	S	O	N	D	cropping intensity actual	cropping intensity potential
wet	maize/sorghum							p	-	-	h			90	100
dry	vegetables	-	-	-	h						p	-	-	20	20
														110	120

3b. Semi-arid/arid savanna (Somalia, Kenya, Southern Sudan)

cropping season	main crops	J	F	M	A	M	J	J	A	S	O	N	D	cropping intensity actual	cropping intensity potential
wet	maize/sorghum				p	-	-	-	h					40	50
wet	cotton				p	-	-	-	-	-	h			30	50
dry	vegetables	-	-	h							p	-	-	30	20
														100	120

4a. Rice - Niger/Senegal rivers

cropping season	main crops	J	F	M	A	M	J	J	A	S	O	N	D	cropping intensity actual	cropping intensity potential
main	rice	-	-	h								p	-	100	100
secondary	rice						p	-	-	-	h			80	100
														180	200

4b. Rice - Gulf of Guinea

cropping season	main crops	J	F	M	A	M	J	J	A	S	O	N	D	cropping intensity actual	cropping intensity potential
main	rice						p	-	-	-	h			100	100
secondary	rice	-	h									p	-	50	100
														150	200

4c. Rice - Southern Sudan

cropping season	main crops	cropping calendar J F M A M J J A S O N D	cropping intensity actual	 potential
I	rice	-------------h p------	100	100
II	rice	p----------------h	80	100
			180	200

4d. Rice - Madagascar tropical lowland

cropping season	main crops	cropping calendar J F M A M J J A S O N D	cropping intensity actual	 potential
main	rice	-------------h p------	100	100
secondary	rice	p----------------------h	30	100
			130	200

4e. Rice - Madagascar highland

cropping season	main crops	cropping calendar J F M A M J J A S O N D	cropping intensity actual	 potential
wet	rice	-------------h p------	100	100
dry	vegetables	p----------------------------------h	10	10
			110	110

5. Egyptian Nile and Delta

cropping season	main crops	cropping calendar J F M A M J J A S O N D	cropping intensity actual	 potential
winter	wheat	-------------------------h p------	40	40
winter	fodder	-------------------h p------------	60	60
summer	maize	p----------------h	50	50
summer	rice	p----------------------h	30	30
			180	180

6. Ethiopian highlands

cropping season	main crops	cropping calendar J F M A M J J A S O N D	cropping intensity actual	 potential
wet	maize	p-----------------------------h	40	70
wet	vegetables	p-----------------------------h	60	30
dry	vegetables	-------------------------------h p------	10	60
			110	160

7. Sudanese Nile area

cropping season	main crops	cropping calendar J F M A M J J A S O N D	cropping intensity actual	 potential
winter	wheat	-------------------h p------	40	40
summer	cotton	p-----------------------------h	50	50
summer	sorghum/maize	p-----------------------------h	40	40
all year	sugar cane	--	10	10
			140	140

8. Shebelli-Juba river area in Somalia

cropping season	main crops	cropping calendar												cropping intensity	
		J	F	M	A	M	J	J	A	S	O	N	D	actual	potential
main/wet	maize				p--	----	----	--h						40	40
secondary	maize	--h									p--	----	----	30	30
main/wet	vegetables				p--	----	----	----	--h					20	20
secondary	vegetables	--h									p--	----	----	30	30
wet	rice				p--	----	----	----	--h					15	15
all year	sugar cane	----	----	----	----	----	----	----	----	----	----	----	----	10	10
														145	145

9. Rwanda - Burundi - Southern Uganda highland

cropping season	main crops	cropping calendar												cropping intensity	
		J	F	M	A	M	J	J	A	S	O	N	D	actual	potential
all year	vegetables/sweet pot	----	----	----	----	----	----	----	----	----	----	----	----	30	30
wet I	maize/sorghum		p--	----	----	----	--h							25	25
wet II	maize/sorghum	--h								p--	----	----	----	15	15
wet I	rice		p--	----	----	----	--h							20	40
wet II	rice	--h								p--	----	----	----	20	50
														110	160

10. Southern Kenya - Northern Tanzania

cropping season	main crops	cropping calendar												cropping intensity	
		J	F	M	A	M	J	J	A	S	O	N	D	actual	potential
wet	vegetables				p--	----	----	----	----	----	--h			40	30
wet	rice				p--	----	----	----	--h					25	35
wet	cotton					p--	----	----	----	----	--h			15	15
all year	sugar cane	----	----	----	----	----	----	----	----	----	----	----	----	10	10
all year	arboriculture	----	----	----	----	----	----	----	----	----	----	----	----	5	5
														95	95

11. Malawi - Mozambique - Southern Tanzania

cropping season	main crops	cropping calendar												cropping intensity	
		J	F	M	A	M	J	J	A	S	O	N	D	actual	potential
wet	rice	----	----	--h								p--	----	40	40
dry	maize				p--	----	----	----	--h					40	40
dry	vegetables				p--	----	----	----	----	----	----	--h		20	20
all year	sugar cane	----	----	----	----	----	----	----	----	----	----	----	----	10	10
														110	110

12a West and Central African humid areas above the equator

cropping season	main crops	cropping calendar												cropping intensity	
		J	F	M	A	M	J	J	A	S	O	N	D	actual	potential
main/wet	rice						p--	----	----	--h				60	70
sec/dry	rice	p--	----	----	----	--h								30	70
dry	vegetables	----	----	----	----	--h							p--	40	20
all year	sugar cane	----	----	----	----	----	----	----	----	----	----	----	----	10	10
														140	170

12b Central African humid areas below the equator

cropping	main	cropping calendar												cropping intensity	
season	crops	J	F	M	A	M	J	J	A	S	O	N	D	actual	potential
main/wet	rice	----	----	----	--h								p--	20	70
sec/dry	rice					p--	----	----	----	--h				10	70
dry	vegetables					p--	----	----	----	----	----	--h		65	20
all year	arboriculture	----	----	----	----	----	----	----	----	----	----	----	----	10	10
														105	170

13. Rivers affluents on Angola - Namibia - Botswana border

cropping	main	cropping calendar												cropping intensity	
season	crops	J	F	M	A	M	J	J	A	S	O	N	D	actual	potential
wet	maize	----	----	----	--h								p--	60	60
wet	vegetables	----	----	----	----	--h						p--	----	40	40
														100	100

14. South Africa - Namibia - Botswana desert and steppe

cropping	main	cropping calendar												cropping intensity	
season	crops	J	F	M	A	M	J	J	A	S	O	N	D	actual	potential
wet	sorghum/maize	----	----	--h									p--	50	50
wet	vegetables	----	----	----	--h								p--	30	30
all year	arboriculture	----	----	----	----	----	----	----	----	----	----	----	----	20	20
														100	100

15. Zimbabwe highland

cropping	main	cropping calendar												cropping intensity	
season	crops	J	F	M	A	M	J	J	A	S	O	N	D	actual	potential
summer	cotton	----	----	----	----	--h						p--	----	30	30
winter	wheat					p--	----	----	----	----	--h			40	40
winter	vegetables					p--	----	----	----	--h				20	20
all year	sugar cane	----	----	----	----	----	----	----	----	----	----	----	----	25	25
														115	115

16. South Africa - Lesotho - Swaziland

cropping	main	cropping calendar												cropping intensity	
season	crops	J	F	M	A	M	J	J	A	S	O	N	D	actual	potential
winter	wheat						p--	----	----	----	----	--h		35	35
wet/sum.	maize	----	----	----	----	--h							p--	25	25
all year	fodder	----	----	----	----	----	----	----	----	----	----	----	----	20	20
all year	pasture	----	----	----	----	----	----	----	----	----	----	----	----	30	30
														110	110

17. Awash river area in Ethiopia

cropping	main	cropping calendar												cropping intensity	
season	crops	J	F	M	A	M	J	J	A	S	O	N	D	actual	potential
wet	cotton						p--	----	----	----	----	----	--h	25	25
wet	maize							p--	----	----	----	--h		25	25
all year	sugar cane	----	----	----	----	----	----	----	----	----	----	----	----	30	30
all year	arboriculture	----	----	----	----	----	----	----	----	----	----	----	----	5	5
														85	85

18. All islands (Comoros, Mauritius, Seychelles, Cape Verde)

cropping season	main crops	cropping calendar												cropping intensity	
		J	F	M	A	M	J	J	A	S	O	N	D	actual	potential
all year	sugar cane	------	------	------	------	------	------	------	------	------	------	------	------	100	100

FIGURE 9
Thiessen polygons for climate stations

TABLE 8

Potential irrigation efficiency and water requirements for the 84 irrigation water requirements zones of Figures 10 and 11

IWR zone	Irrigation crop zone	Irrigation efficiency %	Irrigation water requirement (mm/year)	
			Net	Gross
1	1	60	400	700
2	1	60	500	850
3	1	60	450	750
4	1	60	800	1 350
5	1	60	700	1 200
6	1	60	900	1 500
7	2	70	900	1 300
8	1	60	650	1 100
9	1	60	750	1 250
10	5	80	900	1 150
11	5	80	1 000	1 250
12	5	80	1 250	1 600
13	2	70	900	1 300
14	2	70	1 250	1 800
15	2	70	1 600	2 300
16	2	70	1 200	1 750
17	7	80	1 400	1 750
18	7	80	1 200	1 500
19	7	80	900	1 150
20	7	80	750	950
21	3a	50	400	800
22	3a	50	350	700
23	3a	50	600	1 200
24	3a	50	250	500
25	6	50	350	700
26	17	50	500	1 000
27	3a	50	150	300
28	6	50	100	200
29	6	50	250	500
30	3b	50	1 200	2 400
31	3b	50	750	1 500
32	8	50	750	1 500
33	8	50	450	900
34	3b	50	500	1 000
35	3b	50	850	1 700
36	3b	50	200	400
37	4c	50	1 500	3 000
38	3b	50	350	700
39	10	50	400	800
40	10	50	850	1 700
41	10	50	350	700
42	10	50	600	1 200
43	10	50	350	700
44	10	50	600	1 200
45	11	45	550	1 250
46	9	50	500	1 000
47	9	50	400	800
48	9	50	650	1 300
49	9	50	850	1 700
50	4a	45	2 200	4 900
51	4a	45	1 650	3 700
52	4a	45	1 400	3 150
53	12a	45	1 250	2 800
54	12a	45	900	2 000
55	4b	50	800	1 600
56	12a	45	700	1 600
57	12a	45	550	1 250
58	12a	45	500	1 150
59	12b	45	500	1 150
60	12b	45	550	1 250
61	12b	45	650	1 450
62	12b	45	900	2 000
63	14	65	200	350
64	13	50	300	600
65	14	65	350	550
66	14	65	400	650
67	14	65	600	950
68	16	60	400	700
69	16	60	950	1 600
70	16	60	800	1 350
71	16	60	600	1 000
72	16	60	500	850
73	4d	50	1 000	2 000
74	4e	50	450	900
75	4d	50	750	1 500
76	11	45	500	1 150
77	11	45	550	1 250
78	11	45	450	1 000
79	15	60	650	1 100
80	18	60	300	500
81	18	60	1 500	2 500
82	13	50	200	400
83	18	60	150	250
84	2	70	900	1 300

FIGURE 10
Net irrigation water requirements - Potential scenario after adjustment

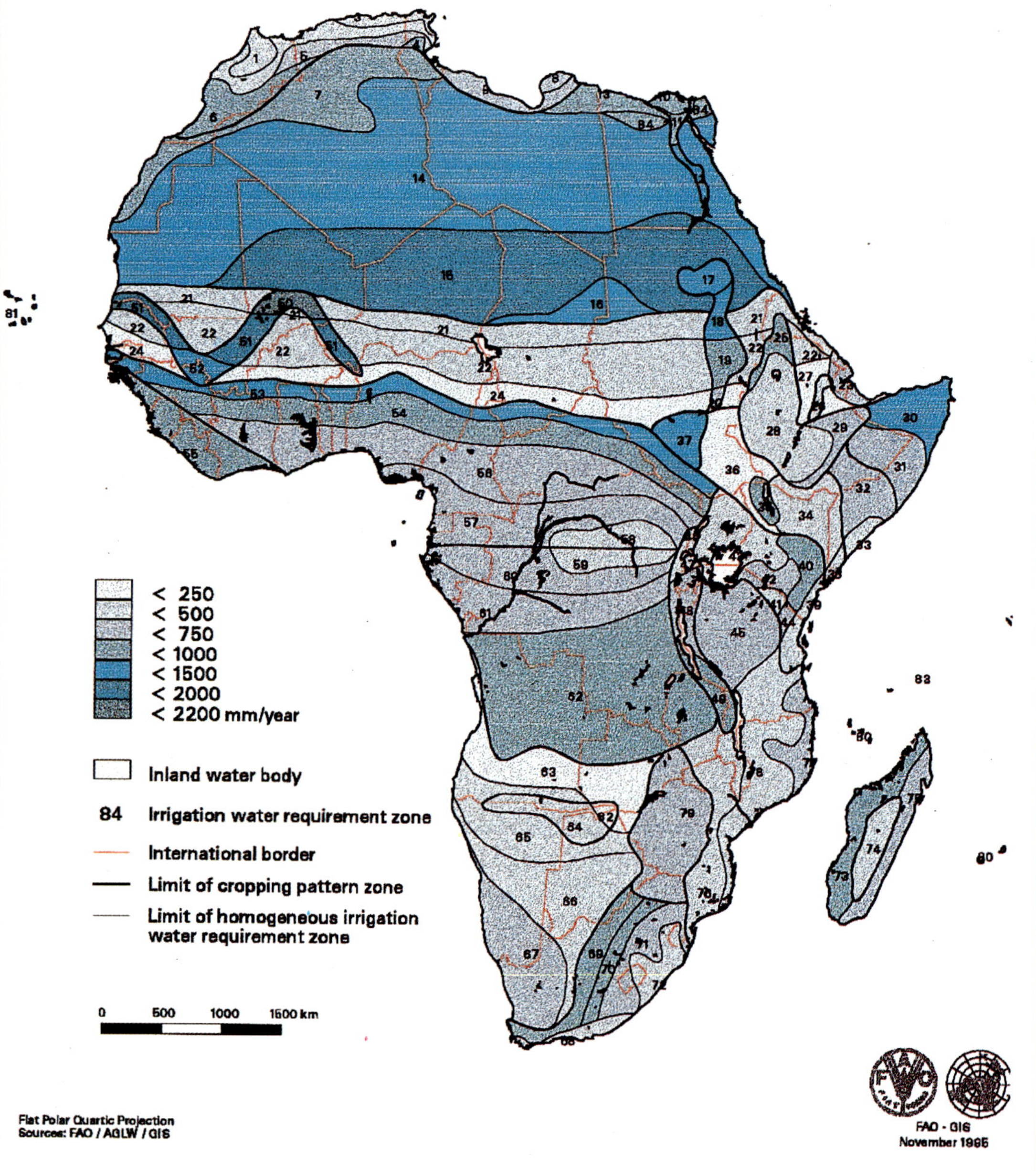

FIGURE 11
Gross irrigation water requirements - Potential scenario after adjustment

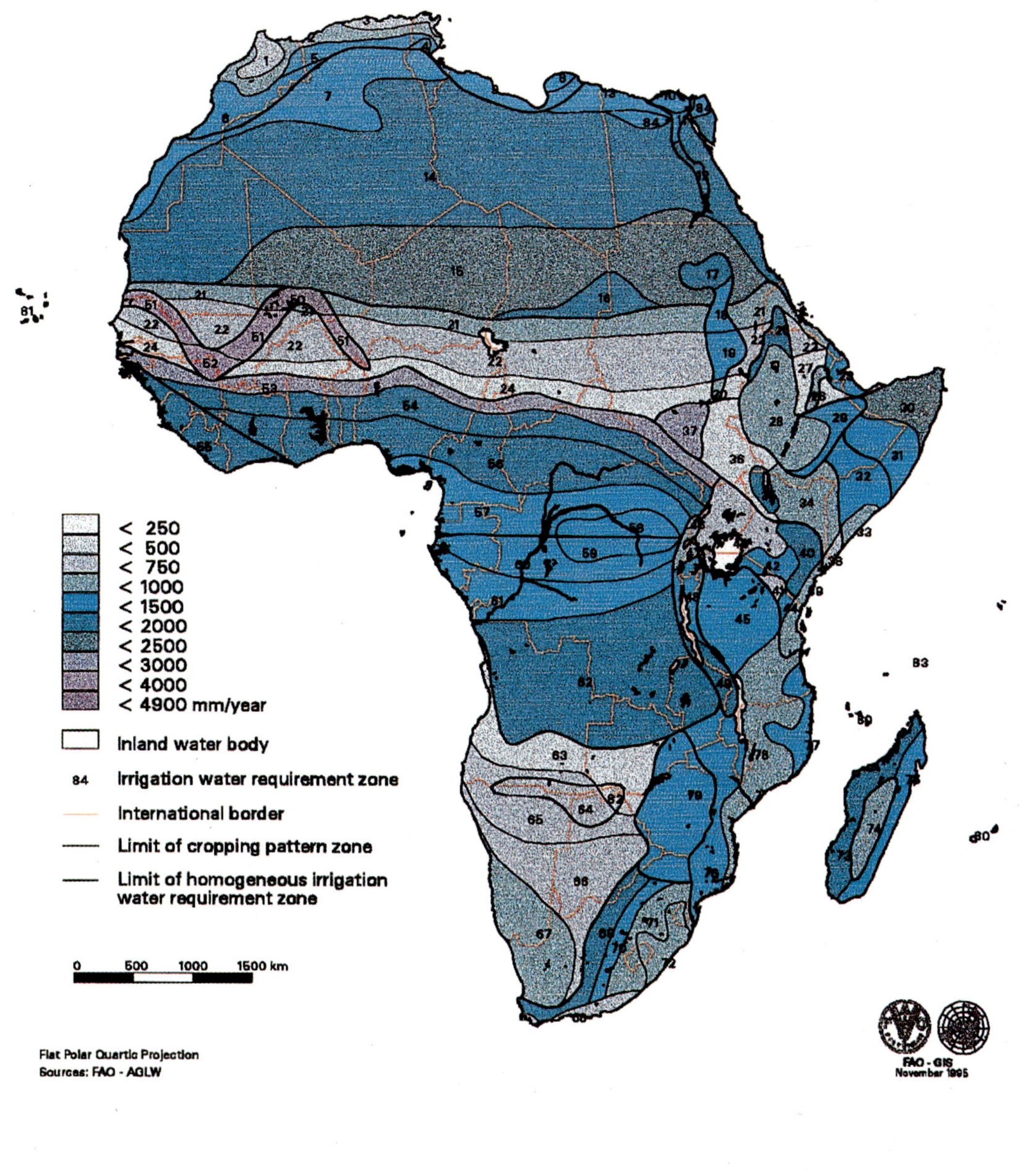

Chapter 6

Review of existing information on irrigation potential

As explained in Chapter 2, the African continent has been divided into 24 major hydrological units, each unit corresponding either to a large river basin or to a group of smaller river basins (Figure 2). The term 'basin', indicating the major hydrological unit, is used in the text and tables of this report.

Table 9 presents some general characteristics of the 24 major basin groups are presented.

TABLE 9
Major basin groups: areas and rainfall (Hutchinson et *al.* 1995 [23])

Basin group number	Name of the basin group	Total area of the basin group (km^2)	As % of the total area of Africa	Annual rainfall in the basin group (mm)		
				min.	max.	mean
01	Senegal River	483 181	1.60	55	2 100	550
02	Niger River	2 273 946	7.51	0	2 845	690
03	Lake Chad	2 381 635	7.86	0	1 590	415
04	Nile	3 112 369	10.28	0	2 060	615
05	Rift Valley	637 593	2.10	90	2 210	650
06	Shebelli-Juba	810 427	2.68	205	1 795	435
07	Congo/Zaire River	3 789 053	12.51	720	2 115	1 470
08	Zambezi	1 351 365	4.46	535	2 220	930
09	Okavango	323 192	1.07	355	1 320	680
10	Limpopo	401 864	1.33	290	1 040	530
11	Orange	896 368	2.96	35	1 040	325
12	South Interior	645 826	2.13	270	905	435
13	North Interior (Sahara)	5 804 463	19.16	0	700	40
14	Mediterranean Coast	679 525	2.24	5	895	235
15	North West Coast	670 621	2.21	0	680	145
16	West Coast	1 430 196	4.72	350	3 395	1 435
17	West Central Coast	704 774	2.33	775	2 830	1 785
18	South West Coast	516 200	1.70	10	1 600	940
19	South Atlantic Coast	365 485	1.21	0	555	190
20	Indian Ocean Coast	663 785	2.19	125	1 770	680
21	East Central Coast	1 026 252	3.39	275	2 305	960
22	North East Coast	725 702	2.40	0	725	165
23	Madagascar	587 040	1.94	400	3 000	1 700
24	Islands	9 346	0.03			
	Total for Africa	**30 290 208**	**100.00**			

The North Interior, which corresponds to the Saharan desert, occupies nearly 20% of the African continent. Rainfall is extremely low in this region, with an annual average of only 40 mm (Figure 7), and the irrigation potential is less than 0.2% of the irrigation potential of the whole continent.

The Congo/Zaire River basin, the West Coast, the West Central Coast and Madagascar are the four wettest regions, with an average annual rainfall of over 1 400 mm, and they also occupy about 20% of the African continent. The sum of the irrigation potentials of these four regions is more than 40% of the irrigation potential of the whole continent. This is 200 times the irrigation potential of the North Interior for same area.

METHODOLOGY AND LIMITATIONS

For each basic unit located within the major basin (see Table 1) all available information was systematically reviewed and cross-checked with the results of the studies of the previous chapters. When discrepancies were found, country based information was generally given precedence over the continental figures.

One difficulty of doing a literature review at continental scale, involving 53 countries and over 1 000 references, is that of inconsistency of information. Although the focus has been on the *physical irrigation potential*, some figures found in the literature might already have also taken into consideration other aspects, such as economic or environmental ones (without, however, explicitly mentioning them). Country studies may implicitly include some assumptions on a reasonable level of investment and demand, and allow for other constraints, like environmental and social factors.

In terms of discharge there is no unique *period of reference*. This can have an important impact on average discharges and thus water availability over different periods. For example, the average annual discharge of the White Nile entering Sudan from Uganda during the period 1961-1980 (50 km^3/year) was nearly twice the average annual discharge during the period 1905-1960 (27 km^3/year). The recent drought years in southern Africa also lead to different averages depending on the period of reference considered. Furthermore, progressive development of agriculture and other water uses reduces discharge and prevents correct assessment of natural flow. All information available on discharges has been reported on the Maps 1 to 22 at the end of this chapter. Discharges are average figures and all figures have a reference. However, as they refer to different periods of reference, they should not be considered as giving a consistent overview of the river discharges of the African continent, but rather as indicative figures.

This review gives no details on *seasonal variations of flows*, which necessitate the construction of storage reservoirs, except for the Nile and the Niger basins which have been studied more in detail. In general such information is available where national water master plans have been drawn up.

Nor does this review give details on the *distance* and *elevation* between suitable land and available water, though irrigation potential figures given by country often take this factor into consideration.

The literature reviewed did not always clearly indicate whether the irrigation potential figure refers to *total potential* or to *identified potential*.

This review concentrates on *surface water resources,* except for arid regions, where the use of groundwater for irrigation purposes already plays an important role, and for the cases where information is readily available. Only *renewable groundwater* was taken into consideration and not the fossil water resources. In general, the priority use of groundwater was considered to be for other purposes (domestic, livestock, etc.), but this report gives no details of other water requirements. Such information is available where national water master plans exist. Oases were not studied in detail, although they may sometimes use renewable water (Mauritania).

For the sake of simplicity, this review considers that if a certain quantity of water is abstracted upstream, the same quantity is subtracted from the resource downstream, except for those basins where a detailed description of the relation between upstream and downstream abstractions is available.

Unsurprisingly, more information is available for arid countries, where water is a limiting factor to agricultural production, than for humid countries, where water is abundant. For those humid countries for which no information was available, estimates or interpolations, based on figures in other similar regions, were combined with results from the GIS study (see chapter 4) to assess the irrigation potential. Where only global figures were available for a country as a whole, the distribution over the different basic units was estimated on the basis of information on land, water and population. Every time the estimate had to be made for one of the two reasons cited above, it is indicated by an asterisk '[*]'.

RESULTS PER MAJOR BASIN GROUP

For each of the 24 major basin groups, a description is given of the main river system and discharges, the irrigation potential, irrigation water requirements and the areas already under irrigation at present. An evaluation of the irrigation potential for each of the 136 basic units as well as for the 53 countries and 24 major basin groups as a whole is given in chapter 8.

The Senegal River basin

The Senegal River basin, located in West Africa, covers 1.6% of the continent and spreads over four countries (Map 1 and Table 10).

TABLE 10
Senegal River basin: areas and rainfall by country

Country	Total area of the country (km^2)	Area of the country within the basin (km^2)	As % of total area of basin (%)	As % of total area of country (%)	Average annual rainfall in the basin area (mm)		
					min.	max.	mean
Guinea	245 857	29 475	6.1	12.0	1 120	2 100	1 475
Mali	1 240 190	139 098	28.8	11.2	455	1 410	855
Mauritania	1 025 520	242 742	50.2	23.7	55	600	270
Senegal	196 720	71 866	14.9	36.5	270	1 340	520
For Senegal basin		483 181	100.0		55	2 100	550

Rivers and discharges

The sources of the Senegal River are located in Guinea and in the wetter south-western part of Mali. Total annual discharge leaving Guinea is estimated at about 8 km^3, but during the dry season the rivers frequently run dry. The Falémé River forms the border between Senegal and Mali over most of its distance. By the time they reach the border point between Mali, Mauritania and Senegal, the different tributaries have become one river, the Senegal River, which then continues to form the border between Senegal and Mauritania. The Karakoro River, flowing into the Senegal River at more or less the same point, originates in Mauritania. The annual discharge of the Senegal River at Bakel is 20 km^3. The Gorgol River, originating in

Mauritania, joins it about 200 km downstream. Further downstream there are no other important tributaries.

Irrigation potential and water requirements

The irrigation potential in **Guinea** is rather limited by the topography. It has been estimated at 5 000 ha [*].

The irrigation potential in **Mali** is also limited by the topography. Once the Manantali dam (on the Bafing tributary in Mali) is operational, it is estimated that about 10 000 ha can be irrigated [182a].

The Senegal River valley in **Mauritania** is rather narrow, with the exception of two depressions in the downstream part. It is expected that with the Manantali dam about 125 000 ha can be irrigated [182a]. In the present transitional period, with the dam not yet fully operational, an artificial flood is created through the dam in order to practise flood recession cropping on an area of 50 000 ha at maximum. In the Gorgol and Karakoro tributary areas the irrigation potential is estimated at a maximum of 40 000 ha, mainly through flood recession cropping with the construction of small earth dams [144]. This brings the total to 165 000 ha. In addition there are some 2 000 ha of oases in the Senegal basin area [145].

For **Senegal**, the prospects for irrigation development in the Falémé basin are very limited: a few hundred hectares. However, with the Manantali dam and the Diama dam (near the mouth of the Senegal River), it is expected that 240 000 ha in the Senegal River valley will be irrigated [181]. During the transitional period the flood created through the Manantali dam allows flood recession cropping on 50 000 ha, as is the case in Mauritania. The Diama dam, essentially designed to prevent intrusion of salt water, was completed in 1985.

TABLE 11
Senegal River basin: irrigation potential, water requirements and areas under irrigation

Country	Irrigation potential (ha)	Gross potential irrigation water requirement		Area under irrigation (ha)
		per ha (m^3/ha per year)	total (km^3/year)	
Guinea	5 000	23 000	0.115	0
Mali	10 000	19 000	0.190	300
Mauritania	165 000	14 000 - 37 000	5.185	46 450
Senegal	240 000	22 000 - 37 000	8.880	71 400
Sum of countries	420 000		14.370	118 150
Total for Senegal basin	< = 420 000		14.370	

If double rice cropping is practised in the Senegal River valley, the water requirements are estimated at 37 000 m^3/ha per year in this study. In the delta, gross water requirements of 18 000 m^3/ha have been measured for rice in the rainy period and well over 20 000 m^3/ha in the dry period [182]. Some literature even gives estimates up to 50 000 m^3/ha per year for double rice cropping [181].

In Guinea and Mali the annual river discharges largely exceed the water requirements for irrigation. However, if double rice cropping is to be considered on the whole potential area in the Senegal River valley and delta in Mauritania and Senegal, the total quantity of water in the basin might not be sufficient to meet the demand for irrigation. Therefore, the figure of 420 000 ha for the basin should be considered as an upper limit.

The Niger River basin

The Niger River basin, located in western Africa, covers 7.5% of the continent and spreads over ten countries (Map 2 and Table 12).

TABLE 12
Niger River basin: areas and rainfall by country

Country	Total area of the country (km^2)	Area of the country within the basin (km^2)	As % of total area of basin (%)	As % of total area of country (%)	Average annual rainfall in the basin area (mm) min.	max.	mean
Guinea	245 857	96 880	4.3	39.4	1 240	2 180	1 635
Côte d'Ivoire	322 462	23 770	1.0	7.4	1 315	1 615	1 465
Mali	1 240 190	578 850	25.5	46.7	45	1 500	440
Burkina Faso	274 000	76 621	3.4	28.0	370	1 280	655
Algeria	2 381 740	193 449	8.5	8.1	0	140	20
Benin	112 620	46 384	2.0	41.2	735	1 255	1 055
Niger	1 267 000	564 211	24.8	44.5	0	880	280
Chad	1 284 000	20 339	0.9	1.6	865	1 195	975
Cameroon	475 440	89 249	3.9	18.8	830	2 365	1 330
Nigeria	923 770	584 193	25.7	63.2	535	2 845	1 185
For Niger basin		2 273 946	100.0		0	2 845	690

Algeria and Chad together cover about 9% of the total Niger River basin, but there are almost no renewable water resources in these areas.

The area of the Niger River basin in Guinea is only 4% of the total area of the basin, but the sources of the Niger River are located in this country. The quantity of water entering Mali from Guinea (40 km^3/yr) is greater than the quantity of water entering Nigeria from Niger (36 km^3/yr), about 1 800 km further downstream. This is due among other reasons to the enormous reduction in runoff in the inner delta in Mali through seepage and evaporation combined with almost no runoff from the whole of the left bank in Mali and Niger.

The most important areas of the Niger basin are located in Mali, Niger and Nigeria (25% in each of these three countries). Mali and Niger are almost entirely dependent on the Niger River for their water resources. In the case of Niger nearly 90% of its total water resources originates outside its borders (the Niger River and other tributaries from Burkina Faso and Benin).

Rivers and discharges

The Niger River, with a total length of about 4 100 km, is the third-longest river in Africa, after the Nile and the Congo/Zaire Rivers, and the longest and largest river in West Africa.

The upper Niger River system

The source of the Niger River farthest away from the mouth is in the mountains of Guinea near the border with Sierra Leone. Together with several tributaries it traverses the interior plateau of Guinea flowing north-east towards the border with Mali. Just after the border it is joined by another tributary which also originates in Guinea. The total annual flow entering Mali from Guinea is estimated at 40 km^3.

The river then proceeds north-east towards the inner delta in Mali, where it is joined at Mopti by an important tributary, the Bani River, which is about 1 100 km long and has its sources in Côte d'Ivoire and Burkina Faso.

The inner delta

The total area covered by the inner delta, which is a network of tributaries, channels, swamps and lakes, can reach about 30 000 km^2 in flood season. The delta area is swampy and the soil sandy. Consequently, the river 'loses' nearly two-thirds of its potential flow between Ségou (at 900 km from its source) and Timbuktu (at 1 500 km) due to seepage and evaporation, the latter being aggravated by the fact that the river here touches the southern flanks of the Sahara desert. All the water from the Bani tributary, which flows into the Niger River at Mopti (at 1 150 km), does not compensate for the 'losses' in the inner delta, as the total flow further downstream still decreases rather than increases (Figure 13). The average 'loss' is estimated at 31 km^3/year, but varies considerably according to the years: it was 46 km^3 during the wet year of 1969 and about 17 km^3 during the dry year of 1973 [29].

FIGURE 12
Average discharges of the Niger River and its main tributaries

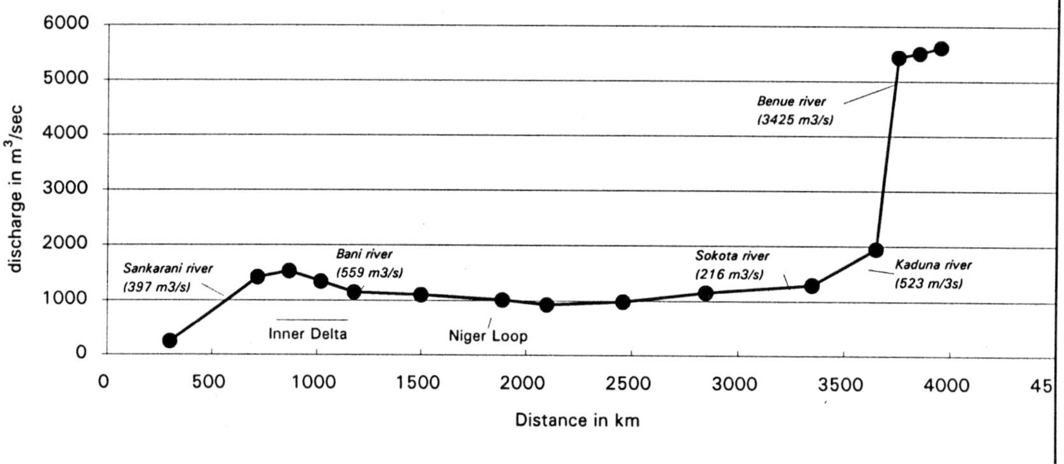

The middle Niger River system

From the inner delta the river continues to flow north-eastwards before turning south-east to form a great bend, the Niger Loop. After meandering through arid areas it enters Niger. In the Niger Loop another 4 km^3/year of water disappear between Diré and Ansongo. Like in the inner delta, these losses are mainly caused by evaporation, but they are much less because of the smaller area inundated during and after the floods. 'Losses' by infiltration are limited.

Within Niger the river receives water from six tributaries originating in Burkina Faso (Gouroual, Dargol, Sirba, Gouroubi, Diamangou, Tapoa). The total annual discharge leaving Burkina Faso is estimated at about 1.4 km^3.

Further downstream the river becomes the border between Niger and Benin, from where three main tributaries enter the river (Mekrou, Alibori, Sota) with a total annual discharge of about 3 km^3.

At Gaya in Niger or Malanville in Benin, just upstream of the border with Nigeria, the average annual discharge has been estimated at about 36 km^3 [35], but only about 18 km^3 was measured in 1986 [29].

The lower Niger River system

Leaving the border between Niger and Benin the river enters Nigeria, where it is joined by numerous tributaries. The most important tributary of the Niger is the Benue which merges with the river at Lokoja in Nigeria. The Benue itself rises in Chad although there are almost no surface water resources in its uppermost part. In Cameroon it receives water from several tributaries. The slope in Cameroon is considerable and the discharge there has important seasonal variations. The quantity of water entering Nigeria was estimated at 25 km^3/year before the 1980s [25] and at 13.5 km^3/year during the 1980s [172]. In Nigeria itself the Benue is joined by several tributaries, of which the ones at the left side originate mainly in Cameroon. The Benue reaches its flood level in September. It begins to fall in October and falls rapidly in November, continuing slowly over the next three months to reach its lowest level in March and April.

From the confluence with the Benue, the Niger heads southwards and empties in the Gulf of Guinea through a network of outlets that constitute its maritime delta.

Table 13 shows the difference between the long term annual flows in Nigeria before the 1980s [30] and the annual flows during the 1980s [172], which was a much drier period.

TABLE 13

Average annual discharges of the Niger River and its main tributaries in Nigeria over different periods

River	Measuring station	Average flow before 1980 (km^3/year)	Average flow in the 1980s (km^3/year)	Difference (%)
Kaduna	Wuya	16.5	14.8	- 10
Benue	Yola	25.0	13.5	- 46
Benue	Makurdi	94.0	74.9	- 20
Benue	Umaisha	108.0	76.7	- 29
Niger	Jebba	40.7	24.3	- 40
Niger	Baro	61.4	43.3	- 29
Niger	Lokoja	171.5	137.9	- 20
Niger	Shintaku	173.8	139.0	- 20
Niger	Idah	177.0	147.3	- 17

Irrigation potential and water requirements

The rainfall and hydrological conditions in **Guinea** make it possible to exploit, with good chances of success for an annual rainfed crop, the alluvial plains of the Niger River and its tributaries. However, to be able to cultivate all year round, irrigation is necessary. The irrigation potential in this region is estimated at 185 000 ha, of which 100 000 ha are relatively easy to develop, though the construction of dams is necessary for the storage of the water [116]. To date only about 6 000 ha of rice are irrigated.

The irrigation potential for the whole of **Côte d'Ivoire** has been evaluated at 475 000 ha, without giving details of location [21a]. It is estimated that 50 000 ha are located in the Niger basin [*].

In **Mali** there are four climate zones in the basin area and rainfall ranges from 1 500 mm in the south to less than 50 mm in the north.

The water in the Niger River is partially regulated through dams. The Sélingué dam on the Sankarani River is mainly used for hydropower, but also permits the irrigation of about 60 000 ha under double cropping [14]. Two diversion dams, one at Sotuba just downstream of Bamako, and one at Markala, just downstream of Ségou, are used to irrigate the area of the Office du Niger (equipped area of about 54 000 ha). However, double cropping in this area would only be possible if the Fomi Dam, planned on the Niandan river in Guinea, were constructed to provide a supplementary and regular amount of water. However, the negative effects on the environment that would be caused by the construction of this dam seem to be important.

Several irrigation projects have been identified, especially related to the construction of the Tala and Djenné Dams on the Bani River and the Dam at Tossaye on the Niger River. However, the drying up of several watercourses during the low-flow period in the dry years 1983-85 requires a careful re-examination of the projects identified, with the recent hydrological figures being taken into consideration [140].

The irrigation potential has been estimated at 556 000 ha, of which about 200 000 ha fully controlled and the rest for partially controlled schemes [138]. At present about 187 000 ha are equipped in the Niger basin, but of this 57 000 ha are already abandoned and of the remaining 130 000 ha actually irrigated more than 60% need to be rehabilitated. Irrigation water requirements for double rice cropping in the Niger River valley range from over 30 000 m^3/ha per year in the south-west to nearly 50 000 m^3/ha per year in the northern part according to this study.

In **Burkina Faso** most of the irrigation is located outside the Niger basin. About 850 ha are irrigated in the Niger basin and the potential is estimated at about 5 000 ha [67].

In **Benin** the irrigation potential has been evaluated at 300 000 ha for the whole country, but no details about location are given [57]. In the present study it has been estimated at 100 000 ha in the Niger basin [*]. The actual equipped area here is 1 090 ha, of which 740 ha are cultivated.

The Niger River crosses the south-western part of **Niger** over a distance of about 550 km with the final 150 km forming the border between Niger and Benin. There are no important tributaries in Niger, but there are two fossil valleys, the Dallols, where there is no permanent flow but where the water resources are quite important. Three other zones are considered as being part of the Niger basin, although in fact they are rather valleys or depressions at a considerable distance from the Niger River with no streams reaching the Niger River: the Ader-Doutchi-Maggia (ADM) valley, the Goulbis valley and the Agadez region.

The total irrigation potential of Niger has been estimated at 222 000 ha, of which 140 000 ha in the Niger River valley and the remaining 82 000 ha spread over the other zones [166]. At present about 54 000 ha benefit from irrigation, of which 16 000 ha are in the Niger River valley.

Irrigation of the 140 000 ha in the Niger River valley and its tributaries on the right side would only be possible through the construction of the Kandadji Dam in the north, just downstream of the border with Mali. Without this dam it would be possible to irrigate only 15 000 ha. However, construction of this multi-purpose dam has so far not been possible due to financial and economic constraints. Reports also indicate that the dam would have a negative impact on the environment [15]. Several other storage works on the tributaries are under consideration.

The irrigation potential in the Niger basin for **Cameroon** has been estimated at 20 000 ha [*]. The Lagdo dam on the Benue River, built primarily for hydroelectricity, regulates the flow of the river. It could also be used for irrigation.

The irrigation sector in **Nigeria** can be divided into three categories [172]:

- public irrigation schemes, which are government-executed schemes;
- farmer-owned and operated irrigation projects (improved fadamas);
- residual fadamas or floodplains.

About 275 000 ha of public schemes are planned under the existing water infrastructure, but only 40 540 ha have been completed and irrigated. As far as the fadamas[1] crop production has depended traditionally on rainfall in the wet season and on residual moisture after flood recession in the dry season. In areas with easily accessible shallow groundwater or surface water, water lifting devices are used to lift water into the land. The existing formal fadama area has been evaluated at 79 000 ha and in addition there are about 550 000 ha of residual fadama cultivation in the Niger basin.

Estimating irrigation potential is rather difficult, despite the considerable data available on surface water resources, because of the potential of large areas to be irrigated either by surface water or shallow fadama aquifers, two sources that are hydraulically connected. Table 14 presents irrigation potential as identified in the national water resources master plan (NWRMP) [172].

TABLE 14
Irrigation potential in the Niger River basin in Nigeria according to the NWRMP [172]

Region in Niger river basin	Potential of public schemes (ha)	Potential of fadama development (ha)	Total irrigation potential (ha)
Niger North	146 590	299 000	445 590
Niger Central	183 140	34 000	217 140
Upper Benue	435 430	320 000	755 430
Lower Benue	61 230	140 000	201 230
Niger South	59 120	0	59 120
TOTAL	885 510	793 000	1 678 510

1 'Fadamas' are sometimes considered wetland, sometimes as flood plains where flood recession cropping is practised.

In this NWRMP there are also two proposals for water transfer schemes from the Niger to the Lake Chad basin and two for water transfer between different tributaries within the Niger basin.

Table 15 summarizes the irrigation potential of the Niger basin, per country and for the basin as a whole.

TABLE 15

Niger River basin: irrigation potential, water requirements, water availability and areas under irrigation

Country with an area within the Niger basin	Irrigation potential	Gross irrigation water requirement		Actual flows		Flows after deduction for irrigation and losses		Area already under irrigation
		per ha	total	inflow	outflow	inflow	outflow	
	(ha)	(m^3/ha.year)	(km^3/yr)	(km^3/yr)	(km^3/yr)	(km^3/yr)	(km^3/yr)	(ha)
	(1)	*(2)*	*(3)*	*(4)*	*(5)*	*(6)*	*(7)*	*(8)*
Guinea	185 000	23 500	4.35	0.00	40.40	0.00	36.05	6 000
Cote d'Ivoire	50 000	23 500	1.18	0.00	5.00	0.00	3.83	0
Mali	556 000	40 000	22.24	45.40	29.20	39.88	6.96	187 500
Burkina Faso	5 000	7 000	0.04	0.00	1.40	0.00	1.37	850
Benin	100 000	18 500	1.85	0.00	3.10	0.00	1.25	740
Niger	222 000	37 000	8.21	33.70	36.30	9.58	3.96	57 520
Cameroon	20 000	18 500	0.37	0.00	13.50	0.00	13.13	2 000
Nigeria	1 678 510	10 000	16.79	49.80	177.00	17.09	rest to sea	670 000
Sum of countries	2 816 510		55.02					924 610
Total for Niger basin	≤ 2 816 510							

NOTES:

For the sake of simplicity it was supposed that if a certain quantity of water is abstracted upstream, this same quantity is subtracted from the resource downstream, except in cases where more information was available.

Mali:

(4) Equal to the sum of the water entering from Guinea (40.40) and Côte d'Ivoire (5.00).

(5) Equal to the water leaving Mali, which is less than the water entering, among others, due to 'losses' in the inner delta.

(6) Equal to the water entering (45.40) minus potential water requirement in Guinea (4.35) and Côte d'Ivoire (1.18).

(7) Equal to the water leaving the country (29.20) minus potential water requirement in Mali (22.24). Potential requirements in Guinea and Côte d'Ivoire are not included, because it is supposed that is is included in the 'losses' in the inner delta. In fact, also a part of the 22.24 km^3 should not be included for this reason.

Niger:

(4)&(5) Outflow (36.30) minus inflow from Mali (29.20), Burkina Faso (1.40) and Benin (3.10) is equal to 2.6 km^3, which is less than the potential water requirement (8.21). In fact Niger needs more water than 'produced' within the country.

(7) Equal to the water leaving the country (36.30) minus potential water requirements in Mali (22.24), Burkina Faso (0.04), Benin (1.85) and Niger (8.21).

Nigeria:

(4) Equal to inflow from Niger (36.30) plus inflow from Cameroon (13.50)

The countries with the largest water requirements are Mali, Niger and Nigeria. Water problems may arise in the Niger basin if the whole potential is developed. The effect of water abstraction upstream of the inner delta on the quantities that disappear within this delta has not been studied. Probably, as is the case with the Sudd swamps in the Nile basin (see section *The Nile basin*), the lower the quantity of water entering the swamp area the lower the quantity of water disappearing in absolute as well as relative terms.

In Nigeria, the most downstream country, of the 177 km^3/year flowing to the sea, only 36 km^3/year enter from Niger and 25 km^3/year from Cameroon. The rest is produced internally. More than 1 million ha of its potential of nearly 1.7 million ha is located in the tributary Benue basin.

In all cases, important storage works for the development of irrigation are necessary throughout the whole basin. Probable navigation and hydropower problems may arise if more water is abstracted for agricultural purposes.

The Lake Chad basin

The Lake Chad basin, located in Northern Central Africa, covers almost 8% of the continent and spreads over seven countries (Map 3 and Table 16).

TABLE 16
Lake Chad basin: areas and rainfall by country

Country	Total area of the country (km^2)	Area of the country within the basin (km^2)	As % of total area of basin (%)	As % of total area of country (%)	Average annual rainfall in the basin area (mm)		
					min.	max.	mean
Nigeria	923 770	179 282	7.5	19.4	285	1 330	670
Niger	1 267 000	691 473	29.0	54.6	0	635	105
Algeria	2 381 740	93 451	3.9	3.9	0	135	20
Sudan	2 505 810	101 048	4.2	4.0	70	1 155	585
Central Africa	622 980	219 410	9.2	35.2	760	1 535	1 215
Chad	1 284 000	1 046 196	43.9	81.5	0	1 350	400
Cameroon	475 440	50 775	2.1	10.7	365	1 590	1 010
For Lake Chad basin		2 381 635	100.0		0	1 590	415

About 20% of the total area of the Lake Chad basin, or 427 500 km^2, is called the Conventional Basin (42% in Chad, 28% in Niger, 21% in Nigeria and 9% in Cameroon), which is under the mandate of the Lake Chad Basin Commission. This commission was created in 1964 by the four member states with the objective of ensuring the most rational use of water, land and other natural resources and to coordinate regional development.

Rivers and discharges

Lake Chad is a terminal depression with the seven basin countries grouped around it, of which four are in direct contact with the lake: Nigeria, Niger, Chad and Cameroon.

In Nigeria, two sub-basins drain into the lake:

- the Yedseram/Ngadda sub-basin to the south;
- the Hadejia/Jama'are-Komadougou/Yobe sub-basin to the north.

The Yedseram River and its tributaries rise in the Mandara hills and it 'loses' most of its water while flowing northwards through a 7-km-wide flood plain. Further downstream, together with the Ngadda River it forms an 80-km^2 swamp and does not maintain a definable water course to the lake.

The Komadougou/Yobe River is the border between Nigeria and Niger over the last 300 km. Upstream of the confluence of the Hadejia and Jama'are rivers the Hadejia-Nguru wetlands (fadamas) start. These cover a total area of about 6 000 km^2 and a water surface area of about 2 000 km^2, but dam construction and increasing water abstraction for irrigation purposes upstream since the 1980s contribute to the fact that large areas of the floodplains are becoming increasingly drier [172]. All rivers crossing this area lose flow as a result of evaporation and evapotranspiration and infiltration to recharge the groundwater. The inflow varies between 1 and 1.8 km^3/year, the outflow between 0.6 and 0.7 km^3/year. When the inflow is more than 2

km^3/year, the outflow gradually increases to 1.2 km^3/year. Upstream the peak flow is at the end of August and rises and falls rapidly reflecting the sporadic nature of heavy rainfalls and the largely impermeable strata. Downstream the peak flow is in January. The flow into Lake Chad is about 0.5 km^3/year. In Niger, in addition to the border Komadougou/Yobe River, there are the Koramas in the south of the country close to the border with Nigeria. These are seasonal rivers and their flow does not reach Lake Chad.

In the north, far away from Lake Chad, is Algeria. The country possesses few renewable water resources. To the east is Sudan with Wadi Kaya and Wadi Azum, both seasonal wadis with spate flows that originate on the western slopes of the Jebel Marra. Their alluvial aquifers could deliver about 0.08 km^3/year of water of excellent quality [30].

To the south is the Central African Republic, a humid country with enormous water resources. The sources of the Chari-Logone Rivers are located in the Central African Republic and the quantity of water leaving the country to Chad was about 33 km^3/year in the period before the 1970s, but fell to 17 km^3/year during the 1980s [29].

The amount of water crossing the border from Cameroon to Chad varies between 3 and 7 km^3/year. More to the north, the Logone River forms the border between Cameroon and Chad until N'Djamena where it flows together with the Chari River which then continues north to the lake. These rivers have a tropical regime with a single flood occurring at the end of the rainy season, which lasts from August to November. They are characterized by irregular inter-annual flows and by their large water 'losses', estimated at about 5 km^3/year, due to flooding of the adjacent Yaéré lowlands in Chad and Cameroon. The largest area flooded covers about 8 000 km^2 and is used for pasture, fishing, flooded rice production and flood recession cropping. In order to expand the Yaéré area, two sites for regulatory dams have been identified on upstream branches of the Logone in Cameroon and Chad. However, this would be to the detriment of water uses for hydro-electric power generation and for irrigation outside these Yaéré lowlands [86].

The rivers outside the Chari-Logone basin in Chad have flash floods during heavy rains and negligible flows the rest of the time, like the Batha River. This regime seriously limits irrigation development.

The Chari-Logone rivers, with 38.5 km^3/year, contribute for about 95% of the total inflow into Lake Chad. In recent history the area of Lake Chad has varied between 3 000 and 25 000 km^2, with a variation in its level of over 8 metres and a variation in volume of between 20 and 100 km^3. The total inflow in recent times has varied between 7 km^3/year (1984/85) and 54 km^3/year (1955/56) [40]. Due to the lowering of the lake level, ideas have been put forward to replenish the lake with water from the Congo/Zaire basin through the construction of a 2 400-km-long canal, but for the time being this is impractical on technical, economic and political grounds [86].

Irrigation potential and water requirements

In **Nigeria**, the planned irrigation under the existing water management works is estimated at 185 000 ha, of which only about 32 000 ha have been completed and irrigated. The total identified potential has been evaluated at 356 000 ha. However, even the complete development of the first 185 000 ha would already create water shortages. In addition, Nigeria plans the development of 146 000 ha of fadamas, of which 20 000 ha in the upper part, 27 000 ha in the middle part and 99 000 ha in the lower part [172].

In **Niger** the irrigation potential in the Koramas sub-basin has been estimated at 8 000 ha, in the downstream Kadougou/Yobe river valley and around the lake at 40 000 ha [167]. In the northern part of the country there are some oases, but no information on them is available.

The irrigation potential in the **Algerian** part of the basin is estimated to be 0 ha [*]. The irrigation potential in **Sudan** is about 4 000 ha [30].

The irrigation potential for the whole of the **Central African Republic** is estimated at 1.9 million ha, but no details are available on location [17]. About one-third of the country is situated in the Lake Chad basin, the remaining two-thirds being in the Congo/Zaire basin. A first approximation of the part of the potential in the Lake Chad basin is estimated at 500 000 ha [*]. This would require 8.25 km^3/year of water, which is about one-quarter to a half of the total quantity of water leaving the country to Chad, depending on the period of reference.

For **Chad**, the irrigation potential has been estimated as follows [21a]:

TABLE 17
Irrigation potential and water requirements in the Lake Chad basin in Chad

Region	Irrigation potential (ha)	Water requirement (km^3/year)
Sudanian and western Sahelian zone:		
- Logone River system	100 000	1.500
- Chari River system	400 000	6.000
- Lake Chad	200 000	3.000
Central and eastern Sahelian zone	135 000	2.025
Total	835 000	12.525

In addition, there are an estimated 90 000 ha of oases in the Saharian zone, but most probably to be irrigated by non-renewable groundwater [85].

The irrigation potential for **Cameroon** is estimated at about 100 000 ha in the Lake Chad basin [*].

Table 18 summarizes the figures for the whole of the Lake Chad basin and for the Conventional Basin.

TABLE 18
Lake Chad basin: irrigation potential and water requirements, result of the country studies

Country	Irrigation potential in whole Lake Chad basin (ha)	Irrigation water requirement (km^3/year)	Irrigation potential in the Conventional Lake Chad basin (ha)	Irrigation water requirement (km^3/year)
Nigeria	502 000	5.020	300 000	3.000
Niger	48 000	0.936	40 000	0.780
Algeria	0	0		
Sudan	4 000	0.030		
Central African Rep.	500 000	8.250		
Chad	835 000	12.525	700 000	10.500
Cameroon	100 000	1.250	80 000	1.000
Total	1 989 000	28.011	1 120 000	15.280

At present, out of a potential of over 1.1 million hectares in the Conventional Basin fewer than 100 000 ha are actually irrigated. However, due to the lowering of the level of Lake Chad in recent history, every new irrigation development has to be studied very carefully. Already in 1980 the maximum development was estimated at fewer than 400 000 ha by a UNDP-financed study [38]. The recently prepared master plan for the Conventional Basin proposes to concentrate future developments on small-scale projects.

Taking into consideration the above aspects, the total potential for the whole of the Lake Chad basin is presented in Table 19.

TABLE 19
Lake Chad basin: irrigation potential, water requirements and areas under irrigation, result of the basin study

Country	Irrigation potential			Gross irrigation water requirement		Area under irrigation
	within conventional basin	outside conventional basin	within the whole basin	per ha	total	
	(ha)	(ha)	(ha)	(m^3/ha.year)	(km^3/year)	(ha)
Nigeria	204 000	100 000	304 000	10 000	3.040	82 821
Niger	3 000	8 000	11 000	19 500	0.215	2 000
Algeria	-	0	0	18 000	0.000	0
Sudan	-	4 000	4 000	7 500	0.030	500
Centr. Afr. Rep.	-	500 000	500 000	16 500	8.250	135
Chad	142 500	135 000	277 500	15 000	4.163	14 020
Cameroon	46 700	20 000	66 700	12 500	0.834	13 820
Total	396 200	767 000	1 163 200		16.531	113 296

The Nile basin

The Nile River, with an estimated length of over 6 800 km, is the longest river flowing from south to north over 35 degrees of latitude. It is fed by two main river systems: the White Nile, with its sources on the Equatorial Lake Plateau (Burundi, Rwanda, Tanzania, Kenya, Zaire and Uganda), and the Blue Nile, with its sources in the Ethiopian highlands. The sources are located in humid regions, with an average rainfall of over 1 000 mm per year. The arid region starts in Sudan, the largest country of Africa, which can be divided into three rainfall zones: the extreme south of the country where rainfall ranges from 1 200 to 1 500 mm per year; the fertile clay-plains where 400 to 800 mm of rain falls annually; and the desert northern third of the country where rainfall averages only 20 mm per year. Further north, in Egypt, precipitation falls to less than 20 mm per year.

The total area of the Nile basin represents 10.3% of the area of the continent and spreads over ten countries (Map 4 and Table 20).

For some countries, like Zaire, the Nile basin forms only a very small part of their territory. Other countries, like Burundi, Rwanda, Uganda, Sudan and Egypt, are almost completely integrated into the Nile basin. However, all the waters in Burundi and Rwanda and more than half the waters in Uganda are produced internally, while most of the water resources of Sudan and Egypt originate outside their borders: 77% of Sudan's and more than 97% of Egypt's water resources as shown in Table 6. Moreover, these latter two countries already use nearly all of the water currently allocated to them, as shown below.

TABLE 20
Nile basin: areas and rainfall by country

Country	Total area of the country (km^2)	Area of the country within the basin (km^2)	As % of total area of basin (%)	As % of total area of country (%)	Average annual rainfall in the basin area (mm)		
					min.	max.	mean
Burundi	27 834	13 260	0.4	47.6	895	1 570	1 110
Rwanda	26 340	19 876	0.6	75.5	840	1 935	1 105
Tanzania	945 090	84 200	2.7	8.9	625	1 630	1 015
Kenya	580 370	46 229	1.5	8.0	505	1 790	1 260
Zaire	2 344 860	22 143	0.7	0.9	875	1 915	1 245
Uganda	235 880	231 366	7.4	98.1	395	2 060	1 140
Ethiopia	1 100 010	365 117	11.7	33.2	205	2 010	1 125
Eritrea	121 890	24 921	0.8	20.4	240	665	520
Sudan	2 505 810	1 978 506	63.6	79.0	0	1 610	500
Egypt	1 001 450	326 751	10.5	32.6	0	120	15
For Nile basin		3 112 369	100.0		0	2 060	615

Rivers and discharges

The most distant source from the sea is the Luvinzora River in Burundi, a tributary of the Kagera River. The Kagera River forms the border between Rwanda and Tanzania, then between Uganda and Tanzania and then flows into Lake Victora, the second-largest freshwater lake in the world with an area of about 67 000 km^2. Total flow into the lake is about 20 km^3/year, of which 7.5 km^3 from the Kagera River, 8.4 km^3/year from the forest slopes in the north-east (Kenya), 3.2 km^3/year from the drier Serengeti Plains in the south-east (Tanzania) and from 1 to 2 km^3/year from the swamps in the north-west (Uganda).

The level of Lake Victoria is extremely sensitive to moderate changes in rainfall over the lake and its tributaries. Average lake rainfall and evaporation are the main factors affecting the lake balance and are more or less equal. As evaporation varies little from year to year, high rainfall gives rise to a disproportionate surplus and also greatly increases the tributary flows which are themselves relatively more variable than the rainfall. The rise in lake level during 1961-64 of about 2 metres seems to be the result of a higher rainfall during that period over the lake and its basin. This surplus then influences the outflow which declines only gradually over a longer period of years [41].

The only outlet of Lake Victoria is at Ripon Falls (Owen Falls Dam) in Uganda. Then begins the Victoria Nile which flows through Lake Kyoga into Lake Albert, also called Lake Mobutu Sesse Seko. This lake also receives water from the Semliki River, which originates in the Mufumbiru mountains in Zaire and flows through Lake Edward to Lake Albert. The combined waters of the Semliki and the Victoria Nile leave Lake Albert at the northern end and become the Albert Nile, which then flows into Sudan.

Uganda is a humid country with numerous lakes and wetlands and with internal renewable water resources globally estimated at 39 km^3/year. However, the total annual flow into the country (at Ripon Falls and from Zaire) is about equal to the total annual outflow to Sudan, which means that a lot of water disappears within the country through evaporation and evapotranspiration from the lakes and wetland.

Entering Sudan, the Albert Nile becomes the Bahr el Jebel. It flows into the Sudd region, the great wetlands which are a maze of channels, lakes and swamps in southern Sudan, and which also receive water from the Bahr el Gazal River, originating in south-west Sudan.

The most remarkable topographic feature of the Sudd area is its flatness: for 400 km, from south to north, the slope is a mere 0.01% and much of it is even flatter. The soils of the whole area are generally clayish and poor in nutrients. Rain falls in a single season, lasting from April to November and varying in the Sudd area from about 900 mm in the south to 800 mm in the north. As the rainy season coincides with, though is slightly shorter than, the flood seasons of the rivers, there is land of water and mud for half of the year and, away from the rivers, land of desert-like dryness for the other half. The main natural channels flow through a swamp area waterlogged throughout the year, and are then flanked by grasslands flooded at high river and exposed when the river level drops. Because of the important rainfall in the Equatorial Lake Plateau during the 1960s and 1970s the permanent swamp area increased from 2 700 km^2 in 1952 to 16 200 km^2 in 1980 [42].

Less than half of the water entering the Sudd region flows out of it into the White Nile. The rest disappears through evaporation and evapotranspiration. The quantity entering the Sudd region varies greatly over the years, mainly depending on the rainfall in the upper catchment area, and hydrological measurements have shown that the greater the flow of water into the Sudd, the greater the percentage of water 'lost' in evaporation (Table 21 [42].

TABLE 21
Average annual discharges at different locations in the Sudd region

Period	Discharge at Mongalla (km^3/year)	Discharge at tail of swamps (km^3/year)	Quantity disappeared (km^3/year)	% disappeared
1905 - 1960	26.8	14.2	12.6	47.0
1961 - 1980	50.3	21.4	28.9	57.5
1905 - 1980	33.0	16.1	16.9	51.2

In order to bypass the Sudd region and to direct downstream a proportion of the water considered lost each year by spill from the river and evaporation in the swamps, the construction of the Jonglei Canal had been planned. This water could then have become available for irrigation and other uses downstream in Sudan and Egypt. Construction of the canal began in 1978 for a planned total length of 360 km, but work stopped in November 1983 after 240 km because of the civil war. By that time it had also become clear that these 'losses' create resources in pasture and fisheries and that the canal causes enormous human and environmental problems in the area. The issue is now how much water can be drained from the Sudd through the construction of the Jonglei Canal without serious and irreparable damage to the local environment and economy and its potential expansion [195].

The Sobat River, that flows into the White Nile just upstream of Malakal, is fed by the Baro and Akobo Rivers and others with catchment areas situated mainly in the southern Ethiopian foothills.

The Blue Nile and its main tributaries, the Dinder and the Rahad, rise in the Ethiopian mountains and around Lake Tana. The confluence of the White Nile and the Blue Nile is at Khartoum. Further downstream is the Atbara tributary, the last important tributary of the Nile system, again deriving from the Ethiopian plateau north-east of Lake Tana and forming the border between Ethiopia and Eritrea before entering Sudan. There are no important tributaries further downstream in Egypt.

The contribution of the rivers of the Ethiopian catchment area (Blue Nile system) to the Nile is about twice the contribution of the rivers of the Equatorial Lake Plateau catchment area (White Nile system), but it is characterized by the extreme range in discharges between the peak and low periods, while the flow from the Equatorial Lake Plateau is more uniform. At its peak the former provides nearly 90% of all water reaching Egypt, the latter only 5%. During the months with low flow the contributions are nearer 30% and 70% respectively [29].

FIGURE 13
Average discharges of the Blue Nile and the White Nile

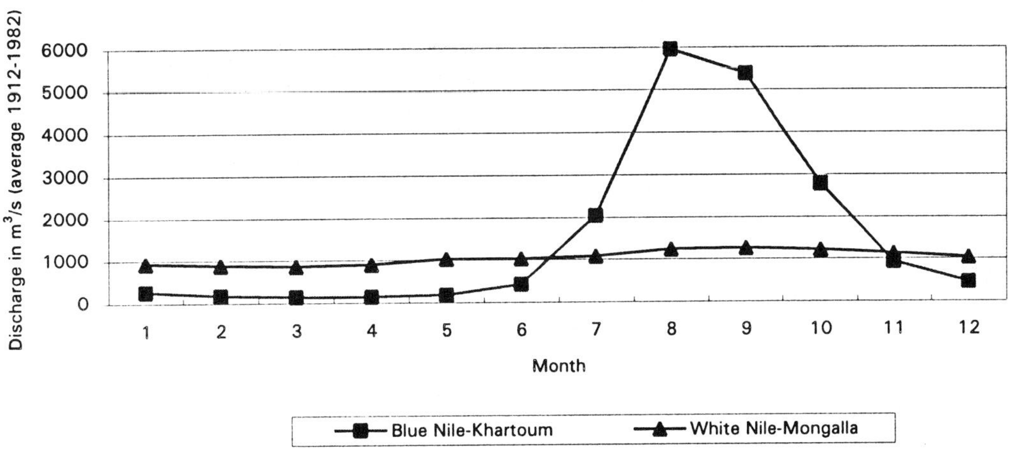

As already mentioned, variations in rainfall over the years can cause quite considerable variations in discharges and lake levels. This seems to be more explicitly the case for the White Nile River system. For this reason, average discharge figures might vary greatly depending on the period under consideration, as shown in Table 22 [29, 210, 44].

TABLE 22
Variations in discharges at different locations on the Nile

Location	Average annual discharges in km^3		
	period 1961-1970	period 1948-1970	period 1912-1982
Lake Victoria exit	41.6	29.4	27.2
Lake Kyoga exit	44.1	30.1	26.4
Lake Albert exit	48.8	33.7	31.4
Mongalla (White Nile)	52.6	36.8	33.1
Malakal (White Nile)	37.8	31.6	29.6
Khartoum (Blue Nile)	45.9	49.8	50.1
Mouth of the Atbara	10.9	12.1	10.6
Dongola (Nile)	86.2	86.2	82.7

In addition to variations due to rainfall, the discharges might vary also due to water abstractions, mainly for irrigation purposes.

Irrigation potential and water requirements

Both **Burundi** and **Rwanda** are characterized by a rolling topography with a continuous pattern of hills and valleys, with lakes and marshy lowlands at the bottom of the valleys. Improving the drainage network in part of the swamp areas, combined where possible with an irrigation network, would allow year-round cultivation, which is important for these small, but very densely populated countries. The total area of these valley bottoms in the Nile basin is estimated at 105 000 ha for Burundi [78] and 150 000 ha for Rwanda [176].

For **Tanzania** the irrigation potential has been estimated at 30 000 ha, but this would require the construction of considerable water conveyance works [199]. In addition to this, at the beginning of the century settlers from Germany, the then colonial power in the country, proposed a plan to transfer water from Lake Victoria to the Vembere Plateau in the Manonga River basin in central Tanzania to irrigate between 88 000 and 230 000 ha of cotton. Though this project is still on the table, it would be very expensive. The transfer would be effected by gravity as the plateau lies below the water level of the lake [199].

The Lake Victoria basin in **Kenya** covers only 8.5% of the total area of the country but it contains over 50% of the national freshwater resources. The national water master plan identified an irrigation potential of 180 000 ha based on 80% dependable flow [125]. As part of the plan, dams and water transfers to other (sub)basins are proposed. At present only about 6 000 ha are irrigated. Moreover, in Kenya there has been lengthy debate as to whether, given adequate technology, Lake Victoria basin water should be transferred to arid areas of the country for irrigation. It is considered that perhaps the most appropriate location for such an experiment would be the Kerio Valley (located in the Rift Valley, see section *The Rift Valley*), for which a special development authority has been established by the Kenyan Parliament. The feasibility of such a project is a question of engineering and several observers consider it possible. Such an undertaking would use significant quantities of water. Projects of this kind are analogous to the irrigation of the Vembere steppe proposed in Tanzania (see above).

The Nile basin in **Zaire** covers less than 1% of the area of the country. The area is hilly and does not really lend itself to irrigation. This area is rather densely populated with most people engaged in cattle rearing and fishery activities around Lake Albert [46]. It is considered that about 10 000 ha could be developed for irrigation [*].

Uganda has large swamp areas covering about 700 000 ha. The irrigation potential is estimated at 202 000 ha, requiring, however, major works such as storage, river regulation and large-scale drainage [209]. At present only 5 550 ha are irrigated.

The irrigation potential in the Nile basin in **Ethiopia** has been estimated at more than 2.2 million hectares [106]. The irrigated area was about 23 000 ha in 1989.

TABLE 23

Water resources, irrigation potential and areas under irrigation in the different Nile sub-basins in Ethiopia

Nile sub-basin	Annual surface runoff (km^3)	Irrigation potential (ha)	Irrigated area in 1989 (ha)
Baro-Akobo	13.4	905 500	350
Blue Nile (Abbay)	54.7	1 001 500	21 010
Setit-Tekeze/Atbara	12.0	312 700	1 800
Total Nile basin	80.1	2 219 700	23 160

The seasonality of the flows in Ethiopia is very high, as shown in Figure 14. This means that very considerable regulation would be necessary for their full utilization. The risk of rapid siltation of the reservoirs because of the steep slopes is a real problem. Construction of dams would augment the quantity of water available, because of a loss of only 3% by evaporation as against a loss of almost 16% in the Aswan reservoir. Egypt, however, would no longer be the beneficiary of additional water in years of high flood, which would then be stored and regulated in the Blue Nile reservoirs instead of Aswan.

The irrigation potential in the Nile basin in **Eritrea** has been estimated at between 60 000 and almost 300 000 ha, though these figures are based on very limited studies [100]. Most of it would be in the Tekeze-Setit basin, which Eritrea shares with Ethiopia. The Mereb-Gash basin has mainly spate flows and its water reaches the Atbara River in Sudan only during extremely high floods. In this review the average irrigation potential has been estimated at 150 000 ha [*].

Irrigation potential in **Sudan** has been estimated at over 4.8 million hectares [193], but this figure does not take into consideration the available water resources. The irrigated area was about 1.6 million hectares in 1979 [195] and 1.9 million hectares in 1990 [196]. There are plans to increase irrigation to about 2.8 million hectares by the year 2000, almost all to be irrigated by Nile water [195].

The figures in Table 24 for irrigated area in 1979 and 1990 correspond to the area equipped for irrigation. The actual irrigated area in 1990 was about 1.2 million hectares, or about 63% of the total equipped area of 1.9 million hectares. About 16.8 km^3 of water was used, corresponding to 14 000 m^3/ha [196]. Despite this relatively high value, water management is a problem, for example water supply on the old established cotton schemes of Gezira-Managil was and is about 12% below crop requirements at crucial points in the growth cycle. At the same time, as much as 30% of the water delivered is not used by crops. In large state-run irrigation projects, like Gezira-Managil and Khashm al Girba, average water deliveries to the command area are between 9 700 and 12 600 m^3 per cultivated hectare per year. Sugar cane, a very water-consuming crop, uses between 28 000 and 40 000 m^3 per ha per year [195].

TABLE 24
Irrigated land use in Sudan [195, 196]

in ha	Available fertile land	Irrigation in 1979	Irrigation in 1990	Planned irrigation in 2000
Nile system:				
White Nile upstream of Malakal	n.a.	16 800	16 800	121 800
White Nile betw. Malakal & Khart.	752 220	209 580	196 140	380 100
Blue Nile upstream of Khartoum	2 633 820	1 132 740	1 270 080	1 525 860
Main Nile betw. Khart. & Egypt	226 800	130 620	147 000	249 060
Atbara	571 200	168 420	168 000	407 820
Mereb-Gash	285 600	n.a.	25 200	> 25 200
Other non-Nilotic streams	372 960	n.a.	29 400	> 29 400
Groundwater	n.a.	n.a.	55 430	> 55 430
Total	4 842 600	1 658 160	1 908 050	> 2 794 670

Considering an availability of 25 km^3 of water for irrigation in 2015 (see Table 25) and a water requirement of 14 000 m^3/ha, only about 1.8 million hectares could be irrigated as opposed to the proposed 2.8 million hectares.

The water balance of Sudan at present, and as proposed over the next 20 years can be summarized as follows [196]:

TABLE 25
Estimated water balance of Sudan in 1995 and 2015 [196]

(in km^3/year)	1995	2015
Water Inputs:		
Sudan share of Nile water (1)	20.55	20.55
Other regional surface runoff	1.45	1.45
Internal runoff	0.70	2.50
Jonglei Canal + swamp reclamation (2)	0.00	4.00
Groundwater	0.70	1.10
Total Water Input	**23.40**	**26.60**
Water Demands:		based on 2% growth/year
Irrigation	16.80	25.00
Domestic	0.80	1.10
Industrial	0.20	0.30
Other (incl reservoir evaporation)	0.20	0.20
Total Water Demand	**18.00**	**26.60**
Net surplus	5.40	0.00

(1) Under the Nile Water Agreement between Sudan and Egypt, the quantity of water allocated to Sudan is 18.5 km^3/year at Aswan, which corresponds to 20.55 km^3 further upstream.

(2) The total amount of water becoming available through the construction of the Jonglei Canal is estimated at 8 km^3 in 2015, of which 50% for Sudan and 50% for Egypt under the agreement between the two countries. Egypt considers 2 km^3 to be available by the year 2000 as shown in its water balance in Table 27. Work on the canal is currently stopped as explained at the beginning of this section.

In **Egypt** the agricultural land use in 1990, almost all irrigated, was as follows [95]:

Table 26
Agricultural land use in Egypt [95]

(in 1 000 ha)	Nile Valley	Nile Delta	New Valley	Coastal Plains	Sinai	Total
Rainfed + supplementary irrigation				126	42	168
Irrigated old lands	798	1 596				2 394
Reclaimed land (pre 1980):						
. cropped	42	210				252
. uncropped	42	42	42			126
Reclaimed (1980-1987):						
. cropped		126				126
. uncropped		84				84
Reclamation (1987-1992)	42	252				294
To be reclaimed by 2000	42	294	84		126	546
Total cropped in 1990	840	1 932		126	42	2 940
Total area including reclamation	966	2 604	126	126	168	3 990

It should be noted that each time new land is reclaimed it is of a lower quality than the already cultivated land. The best soils in Egypt cover an area of only about 1 million ha [20], while the best plus suitable soils cover an area of about 3.6 million ha. Adding the still more marginal land, the maximum area for agriculture could be 4.8 million ha [20]. The remaining soils are unsuitable for agriculture.

Taking into consideration water saving and possibilities of re-use, the water balance of the Nile basin in Egypt in 1993 and 2000 is presented in Table 27.

Taking an average water requirement of 13 000 m^3/ha per year in the Nile Valley and Delta in this study, about 4 420 000 ha could be irrigated using the 57.4 km^3/year of Nile water.

As can be seen from Table 28, the sum of the irrigation potential of the countries leads to a water deficit of over 26 km^3/year, (column 7) without considering possibilities of re-using water as indicated by Egypt and Sudan in their water balance, but after deducting the water 'losses' in the Sudd region.

This deficit corresponds to an area of almost 2.2 million hectares, considering an average water requirement in the region of 12 000 m^3/ha per year [*]. This leads to an irrigation potential for the basin as a whole of 8 million hectares instead of the nearly 10.2 million hectares. However, even these 8 million hectares are still a very optimistic estimate and should be considered as a maximum value, requiring very important storage works and optimum water use.

TABLE 27

Estimated water balance of Egypt in 1993 and 2000

(in km^3/year)	1993	2000
Water Inputs:		
Surface water resources (1)	56.0	58.0
Groundwater in Nile Valley and Delta	2.3	4.8
Agricultural drainage water	4.0	6.5
Treated sewage water	0.2	1.2
Improved water management	0.0	1.0
Total Water Input	**62.5**	**71.5**
Water Demands:		
Irrigation	47.4	57.4
Municipal	3.1	3.1
Industrial	4.6	6.1
Navigation, etc.	1.8	0.3
Total Water Demand	**56.9**	**66.9**
Net Surplus	5.6	4.6

(1) It is expected that the first phase of the construction of the Jonglei Canal will be terminated by 2000, giving 2 km^3 per year of water both to Sudan and to Egypt.

TABLE 28

Nile basin: irrigation potential, water requirements, water availability and areas under irrigation

Country area within the Nile basin	Irrigation potential	Gross irrigation water requirement		Actual flows		Flows after deduction for irrigation and losses		Area already under irrigation
		per ha	total	inflow	outflow	inflow	outflow	
	(ha)	(m^3/ha.year)	(km^3/yr)	(km^3/yr)	(km^3/yr)	(km^3/yr)	(km^3/yr)	(ha)
	(1)	*(2)*	*(3)*	*(4)*	*(5)*	*(6)*	*(7)*	*(8)*
Burundi	80 000	13 000	1.04	0.00	1.50	0.00	0.46	0
Rwanda	150 000	12 500	1.88	1.50	7.00	0.46	4.09	2 000
Tanzania	30 000	11 000	0.33	7.00	10.70	4.09	7.46	10 000
Kenya	180 000	8 500	1.53	0.00	8.40	0.00	6.87	6 000
Zaire	10 000	10 000	0.10	0.00	1.50	0.00	1.40	0
Uganda	202 000	8 000	1.62	28.70	37.00	23.83	30.51	9 120
Ethiopia	2 220 000	9 000	19.98	0.00	80.10	0.00	60.12	23 160
Eritrea	150 000	11 000	1.65	0.00	2.20	0.00	0:55	15 124
Sudan	2 750 000	14 000	38.50	117.10	55.50	90.63	31.13	1 935 200
Egypt	4 420 000	13 000	57.46	55.50	rest to se	31.13	minus26.33	3 078 000
Sum of countries	10 192 000		124.08					5 078 604
Total for Nile basin	< 8 000 000							

NOTES:

For the sake of simplicity it was supposed that if a certain quantity of water is abstracted upstream, this same quantity is subtracted from the resource downstream, except in cases where more information was available.

Tanzania:

(6) Equal to inflow (7.00) minus water requirement upstream countries (1.04 + 1.88).

(7) Equal to outflow (10.70) minus water requirement upstream and within Tanzania (1.04 + 1.88 + 0.33).

Uganda:

(6) Equal to inflow (28.70) minus water requirement upstream countries (1.04 + 1.88 + 0.33 + 1.53 + 0.10).

(7) Equal to outflow (37.00) minus water requirement upstream and within Uganda (1.04 + 1.88 + 0.33 + 1.53 + 0.10 + 1.62).

Sudan:

(1) Not included the possibility of irrigation within the Sudd area (area about 1 600 000 ha).

(4) Total inflow from Uganda and Ethiopia.

(5) Attribution to Egypt according to 1959 agreement after deduction evaporation Aswan.

(6) Equal to inflow (117.1) minus water requirement equatorial plateau countries (6.50) and Ethiopia (19.98).

(7) Equal to outflow (90.63) minus losses in Sudd (21) and water requirement within Sudan (38.50).

Egypt:

(4) Attribution to Egypt according to 1959 agreement after deduction evaporation Aswan.

(6) Equal to outflow from Sudan after potential deductions (31.13) minus water requirements (57.46).

The Rift Valley

The Rift Valley, located in Eastern Africa, covers just over 2% of the continent and spreads over seven countries (Map 5 and Table 29).

TABLE 29
The Rift Valley: areas and rainfall by country

Country	Total area of the country (km^2)	Area of the country within the basin (km^2)	As % of total area of basin (%)	As % of total area of country (%)	Average annual rainfall in the basin area (mm) min.	max.	mean
Djibouti	23 200	12 800	2.0	55.2	110	345	155
Eritrea	121 890	8 605	1.3	7.1	95	545	230
Ethiopia	1 100 010	310 981	48.8	28.3	90	1 990	725
Sudan	2 505 810	16 441	2.6	0.7	360	1 320	515
Uganda	235 880	4 514	0.7	1.9	385	1 540	710
Kenya	580 370	130 452	20.5	22.5	155	1 545	480
Tanzania	945 090	153 800	24.1	16.3	370	2 210	690
For Rift Valley		637 593	100.0		90	2 210	650

The Rift Valley consists of a group of independent interior basins, extending from Djibouti in the north to Tanzania in the south, nearly half being located in Ethiopia.

TABLE 30
The different basins within the Rift Valley

Name of basin		Total area of basin (km^2)	Area in the country (km^2)
Danakil:		92 741	
	Djibouti		11 800
	Eritrea		8 605
	Ethiopia		72 336
Awash:		112 030	
	Djibouti		1 000
	Ethiopia		111 030
Central lakes:		54 070	
	Ethiopia		51 070
	Kenya		3 000
Omo-Gibe:		199 952	
	Ethiopia		76 545
	Sudan		16 441
	Uganda		4 514
	Kenya		102 452
Southern lakes:		178 800	
	Kenya		25 000
	Tanzania		153 800
Total		637 593	637 593

Rivers and discharges

The **Danakil basin** is a very dry basin and only rainfall of more than 10 mm results in rapid floods lasting not more than a few hours. Annual runoff is less than 1 km^3.

Lake Abbé, a salt lake on the border between Djibouti and Ethiopia, is in the **Awash basin**. The main part of the Awash basin is in Ethiopia, with annual rainfall ranging from 200 mm in the north to over 1 900 mm in the south. The annual runoff in this basin is estimated at 4.6 km^3 [108].

The **Central lakes basin**, which groups several lakes, is also mainly located in Ethiopia, with a small part continuing into Kenya. Total annual runoff is estimated at 5.64 km^3 [108].

The **Omo-Gibe basin**, with rivers flowing into Lake Turkana (also called Lake Rudolph) is mainly located in Ethiopia and Kenya, with small parts in Sudan and Uganda. From Ethiopia the Omo and Gibe Rivers flow into the lake, while from Kenya the Turkwel and Kerio Rivers flow into the lake. Annual runoff in this basin is estimated at 16.1 km^3 [108].

In the southern part of Kenya and the northern part of Tanzania the **Southern Lakes** basins are grouped, of which Lake Natron and Lake Eyasi are the most important ones.

Irrigation potential and water requirements

Agriculture in **Djibouti** is only possible with irrigation. The cultivable land is estimated at 6 000 ha, but only 674 ha are equipped for irrigation, of which 300 ha are in the Rift Valley [93]. No detailed information is available on irrigation potential, but with the available water resources it has been estimated at 1 000 ha, of which 450 ha are estimated to be in the Rift Valley [*].

A narrow strip along the south-eastern border of **Eritrea** drains into the Danakil depression. Due to its closed topography and arid climate, it is characterized by highly saline soils and groundwater and has little agricultural potential. The potential for irrigation is estimated to be negligeable [*].

Most of the irrigation developed to date in **Ethiopia** is located in the Awash basin. The irrigation potential in the Rift Valley region in Ethiopia is estimated at 790 000 ha, distributed over the different basins as follows [106]:

TABLE 31
Water resources, irrigation potential and water requirements in the different Rift Valley basins in Ethiopia [106]

Basin in Ethiopia	Annual runoff (km^3)	Irrigation potential (ha)	Gross water requirement (m^3/ha.year)	Annual water requirement (km^3)
Danakil	0.86	0	5 000	0
Awash	4. .60	205 400	10 000	2.05
Central Lakes	5.64	139 300	9 000	1.25
Omo-Gibe	16.10	445 300	9 000	4.01
	27.20	790 000		7.31

While the total water requirement is only one-fourth of the annual runoff, the development of the irrigation potential would require important storage works.

Less than 1% of **Sudan** lies in the Rift Valley. It is a swampy area, with no information available on water resources and irrigation potential.

The border between Uganda and Kenya coincides more or less with the watershed line of the Rift Valley basin. In **Uganda** the water resources are rather limited. The irrigation potential is estimated to be negligible [*].

The difference in climate in the Rift Valley in **Kenya** is quite distinct. The rainfall is considerable, more than 1 500 mm/year at the edges of the Rift Valley and decreasing rapidly to under 200 mm in the valley bottom. In this study, irrigation water requirements are estimated at 10 500 m^3/ha per year in the north and at 12 000 m^3/ha per year in the south. The irrigation potential, identified in the national water master plan and based on 80% dependable flow, is estimated at about 35 900 ha in the northern Omo-Gibe basin and 16 600 ha in the southern Lower Ewaso Ng'iro basin [125]. For the Kerio Valley, located in the Omo-Gibe basin, a special development authority has been established by the Kenyan Parliament to study the possibility of transferring water from the Lake Victoria basin to this basin (as explained in the section *The Nile basin*).

For **Tanzania** an irrigation potential of only 1 060 ha has been identified in this area [199]. This basin is also the location of the Vembere Plateau of the Manonga River basin, an area for which plans to transfer water from the Lake Victoria basin have existed since the beginning of the century, as explained in the section *The Nile basin*.

Table 32 summarizes the irrigation potential and the water requirements for the whole Rift Valley region.

TABLE 32
Rift Valley: irrigation potential, water requirements and areas under irrigation

Country	Irrigation potential (ha)	Gross potential irrigation water requirement		Area under irrigation (ha)
		per ha (m^3/ha per year)	total (km^3/year)	
Djibouti	450	12 000	0.005	100
Eritrea	0	8 000	0.000	0
Ethiopia	790 000	5 000 - 10 000	7.315	166 396
Sudan	0	7 000	0.000	0
Uganda	0	5 500	0.000	0
Kenya	52 500	10 500 - 12 000	0.576	27 000
Tanzania	1 060	12 000	0.013	0
Sum of countries	844 010		7.910	193 496
Total for Rift Valley	844 010		7.910	

No water problems are expected for the development of this potential, though a lot of storage works will be necessary.

The Shebelli-Juba basin

This basin occupies about one-third of Ethiopia, one-third of Kenya and one-third of Somalia and covers about 2.7% of the continent (Map 6 and Table 33).

TABLE 33
Shebelli-Juba basin: areas and rainfall by country

Country	Total area of the country (km^2)	Area of the country within the basin (km^2)	As % of total area of basin (%)	As % of total area of country (%)	Average annual rainfall in the basin area (mm)		
					min.	max.	mean
Ethiopia	1 100 010	373 739	46.1	34.0	220	1 470	490
Kenya	580 370	210 226	25.9	36.2	205	1 795	395
Somalia	637 660	226 462	27.9	35.5	250	585	375
For She-Jub basin		810 427	100.0		205	1 795	435

Rivers and discharges

The Shebelli and Juba Rivers originate in Ethiopia and flow together just before the mouth in Somalia.

Over 90% of the discharge of the Shebelli River originates from runoff in the Ethiopian highlands and there are large inter-annual variations in discharge. The surface water resources in Ethiopia are estimated at 3.2 km^3/year. Within Somalia the discharge decreases rapidly as it flows to its confluence with the Juba River, as a result of losses by seepage, evaporation and overbank spillage due to a low channel capacity [186]. Often the river ceases to flow in the lower reaches during the early part of the year.

The water resources of the Juba River in Ethiopia are estimated at 5.9 km^3/year. The river crosses Somalia for a distance of 875 km and is one of the important rivers of east Africa. Within Somalia its discharge decreases significantly for the same reasons as the Shebelli River. This river can also cease to flow in the early part of the year. While the basin area of the Juba River at the border with Ethiopia is smaller than that of the Shebelli River, its discharge is almost three times as much due to geological conditions.

The part of the basin in Kenya collects drainage from the northern side of Mount Kenya and the Aberdares, and from smaller mountains or uplands in the north and north-east. Except for the Ewaso Ng'iro River itself, streams flow only in direct response to rainfall. The water reaches the border with Somalia only in very wet years.

Irrigation potential and water requirements

The irrigation potential in the Shebelli basin in **Ethiopia** has been estimated at 204 000 ha [106]. Considering an irrigation water requirement in this region of 14 000 m^3/ha per year in the present study, this would lead to a total annual irrigation water requirement of over 2.8 km^3, which is already more than the water resources available for agriculture, estimated at 2.5 km^3/year [103]. The irrigation potential in the Juba basin has been estimated at 423 300 ha [106], requiring nearly 6 km^3/year of water, which is also more than the 4 km^3/year estimated to be available for agriculture [103].

In **Kenya** sufficient water is available in the upper basin, but no significant storage sites can be located to control the natural flow. In the lower part some suitable sites are available. According to the national water master plan 9 460 ha could be irrigated, based on 80% dependable flow [125].

In **Somalia** it is estimated that, if the flow could be regulated, 60 000 ha could be irrigated in the Shebelli basin [188]. In the Juba basin, the planned, but up to now never constructed Baardhere dam was designed to irrigate up to 170 000 ha, but the size of the dam already seems to have been reduced to irrigate 50 000 ha, in view of the sharing of water with Ethiopia [103].

TABLE 34

Shebelli-Juba basin: irrigation potential, water requirements and areas under irrigation

Country	Irrigation potential (ha)	Gross potential irrigation water requirement		Area under irrigation (ha)
		per ha (m^3/ha per year)	total (km^3/year)	
Ethiopia	627 300	14 000	8.782	0
Kenya	9 460	11 000	0.104	0
Somalia	230 000	14 000	3.220	199 000
Sum of countries	866 760		12.106	199 000
Total for She-Juba basin	≤ 351 460		5.000	

In view of the total available water resources, it will not be possible to irrigate all the areas proposed by Ethiopia and Somalia. If 5 km^3/year of water is available for agricultural purposes, the total irrigation potential has to be reduced by 60% to about 350 000 ha.

The Congo/Zaire River basin

This basin is the largest river basin of Africa, covering over 12% of the continent. It extends over nine countries and the largest area is in Zaire (Map 7 and Table 35). It is one of the most humid basins of Africa.

Rivers and discharges

Its sources farthest away from the mouth are located in Zambia, one draining to Lake Tanganyika, estimated at 2 km^3/year, and one to Lake Mweru, where the flow at the outlet is estimated at over 41 km^3/year. No information is available about the sources originating in Tanzania and flowing into Lake Tanganyika. The flows in Burundi drain mainly into Lake

TABLE 35
The Congo/Zaire River basin: areas and rainfall by country

Country	Total area of the country (km²)	Area of the country within the basin (km²)	As % of total area of the basin (%)	As % of total area of the country (%)	Average annual rainfall in the basin area (mm)		
					min.	max.	mean
Zambia	752 610	177 735	4.7	23.6	985	1 420	1 195
Tanzania	945 090	244 593	6.5	25.9	720	1 385	970
Burundi	27 834	14 574	0.4	52.4	920	1 565	1 155
Rwanda	26 340	6 464	0.2	24.5	1 135	1 580	1 365
Central Africa	622 980	403 570	10.7	64.8	1 065	1 680	1 465
Cameroon	475 440	96 395	2.5	20.3	1 440	1 670	1 545
Congo	342 000	246 977	6.5	72.2	1 190	1 990	1 660
Angola	1 246 700	285 395	7.5	22.9	785	1 635	1 375
Zaire	2 344 860	2 313 350	61.1	98.7	775	2 115	1 540
For Congo/Zaire basin		3 789 053	100.0		720	2 115	1 470

Tanganyika and those in Rwanda into Lake Kivu, which is connected with Lake Tanganyika through the Rusizi border river between Zaire, Rwanda and Burundi.

In the north about two-thirds of the Central African Republic lie within the Congo/Zaire basin. It is a humid country, with many sources flowing into the Oubangui River, a major tributary of the Congo/Zaire River and forming the border between the Central African Republic and Zaire. At Bangui, its discharge is estimated at over 126 km^3/year. The tributaries originating within Cameroon flow either to the Central African Republic in the east or to Congo in the south, where the discharge of the Sangha River at the border is over 52 km^3/year. The Oubangui tributary forms the border between Congo and Zaire, then flows into the Congo/Zaire River which continues to be the border until the far south-west where it enters Zaire. Many other tributaries originate in Congo.

To the south is Angola, where the Kasai River, another major tributary, originates together with many other smaller tributaries.

Zaire has a very dense hydrographic system (Figure 6). The discharge of the Congo/Zaire River reaching Kinshasa and Brazzaville is about 1 269 km^3/year, which is equal to 32% of the renewable water resources for the whole of Africa. The river then continues to the south-west and forms the border between Angola and Zaire before flowing into the sea.

Irrigation potential and water requirements

It is difficult to find reliable estimates of the irrigation potential of the very humid countries, like Zaire, the Central African Republic, Congo or Angola. In fact, neither water nor land is a limiting factor to agricultural development in these countries and other factors have to be taken into account in order to have some kind of realistic estimates of potential. Methods for assessing irrigation potential in these countries are described below.

For **Zambia** a national water master plan exists [212]. The identified irrigation potential in the Congo/Zaire basin in Zambia has been estimated at 101 000 ha, of which 15 000 ha by renewable groundwater and 20 000 ha of wetlands (dambos).

There is no detailed information on the irrigation potential in **Tanzania**. The Luichi Delta near Kigoma on the shores of Lake Tanganyika contains a large area of good land which is seasonally flooded and unusable. Reclamation of 3 000 ha has been proposed by means of a flood control dam and improved drainage. No irrigation has been proposed in this area [199].

For **Burundi**, about half of which is located in the Congo/Zaire basin, the irrigation potential has been estimated at about 105 000 ha in the basin, of which 75 000 ha for fully or partially controlled irrigation in the plains and the remaining area consisting of valley bottoms (bas-fonds) [78]. About 25% of **Rwanda** is in the Congo/Zaire basin and the irrigation potential here is estimated at 9 000 ha, mainly consisting of valley bottoms [176].

Of the total irrigation potential of the **Central African Republic**, evaluated at 1.9 million hectares [50], about 1.4 million hectares are estimated to be within the Congo/Zaire basin [*]. For **Cameroon** 50 000 ha are estimated to be in the Congo/Zaire basin [*]. For **Congo** the potential is estimated at between 40 000 and 340 000 ha [60, 50]. In this study, the rather abitrary upper estimate has been retained and 255 000 ha are considered to be located in the Congo/Zaire basin, which occupies 75% of the country [*].

For **Angola** a figure of 6.7 million hectares exists for irrigation potential [17]. According to the Direction of the Hydraulical Service about 420 000 ha could be irrigated at present, considering land and water as well as human resources [51]. The present study has evaluated the irrigation potential at 3% of the area of the country, which corresponds to 3.7 million hectares. It has been distributed over the six basins on the basis of the percentage of the country covered by each basin, except for the Okavango and the South Interior, which are less humid than the other ones [*]:

TABLE 36
Irrigation potential estimates in the different major basin groups in Angola [*]

Name of basin	Estimated potential in ha [*]
Congo/Zaire	900 000
Zambezi	700 000
Okavango	200 000
Souh Interior	50 000
West Central Coast	50 000
South West Coast	1 800 000
Total	3 700 000

Almost 99% of **Zaire** is located within the Congo/Zaire basin. About 75% of the country, which is over 170 million hectares, is covered by natural forest, most of it untouched. Of these 170 million hectares, about 139 million hectares are considered exploitable and capable of producing 700 million m^3 of industrial wood per year (or 5 m^3 per ha). At present only 0.5 million m^3 per year is produced due mainly to infrastructure and transport problems. Of the remaining 63 million hectares, only 6 million ha is cultivated. Irrigation potential figures vary between 4 and 20 million hectares [32, 21a]. While it is true that the water resources of the country are abundant, it is not realistic to estimate that 20 million ha can be developed for irrigation, which is more than three times the total area cultivated at present! Land suitable for agriculture has been estimated at 80 million ha. When considering that no forest land will be converted into agricultural land, this area is reduced to 60 million ha, about half of which is used for other purposes. Like for Angola, this study estimates the irrigation potential at 3% of the total area of the country, or 7 million ha [*]. This area requires 108.50 km^3/year of water for irrigation, which is about 12 % of the internal renewable water resources of Zaire, estimated at 935 km^3/year. Of these 7 million ha, 10 000 ha are considered to be in the Nile basin and 10 000 ha in the West Central Coast region. The remaining area, 6.98 million hectares, is considered to be in the Congo/Zaire basin.

Unlike the drier basins, where the irrigation potential figures given in this study should be considered as a maximum from the point of view of water resources, the figures for humid basins where water is abundant are rather arbitrary.

TABLE 37
Congo/Zaire basin: irrigation potential, water requirements and areas under irrigation

Country	Irrigation potential (ha)	Gross potential irrigation water requirements per ha (m^3/ha per year)	Gross potential irrigation water requirements total (km^3/year)	Area under irrigation (ha)
Zambia	101 000	19 500	1.970	5 000
Tanzania	0	13 000	0.000	0
Burundi	105 000	13 000	1.365	14 400
Rwanda	9 000	13 000	0.117	2 000
Central Africa	1 400 000	18 000	25.200	0
Cameroon	50 000	14 000	0.700	1 650
Congo	255 000	13 000	3.315	217
Angola	900 000	20 000	18.000	2 000
Zaire	6 980 000	15 500	108.190	10 500
Sum of countries	9 800 000		158.857	35 767
Total for Congo/Zaire	9 800 000		158.857	

The Zambezi basin

The Zambezi basin is the fourth-largest river basin of Africa, after the Congo/Zaire, Nile and Niger basins. Its total area represents about 4.5% of the area of the continent and spreads over eight countries (Map 8 and Table 38). The Zambezi River flows eastwards for about 3 000 km from its sources to the Indian Ocean.

TABLE 38
The Zambezi basin: areas and rainfall by country

Country	Total area of the country (km^2)	Area of the country within the basin (km^2)	As % of total area of basin (%)	As % of total area of country (%)	Average annual rainfall in the basin area (mm) min.	max.	mean
Angola	1 246 700	235 423	17.4	18.9	550	1 475	1 050
Namibia	824 900	17 426	1.3	2.1	545	690	630
Botswana	581 730	12 401	0.9	2.1	555	665	595
Zimbabwe	390 760	213 036	15.8	54.5	535	1 590	710
Zambia	752 610	574 875	42.5	76.4	600	1 435	955
Tanzania	945 090	27 840	2.1	2.9	1 015	1 785	1 240
Malawi	118 480	108 360	8.0	91.5	745	2 220	990
Mozambique	801 590	162 004	12.0	20.2	555	1 790	905
For Zambezi basin		1 351 365	100.0		535	2 220	930

Rivers and discharges

The Zambezi River rises in the Kalene hills in north-western Zambia and flows northwards for about 30 km. It then turns west and south to run over about 280 km through Angola and re-enters Zambia with an annual discharge of nearly 18 km^3. It then flows southwards through marshy plains. In the south-west of Zambia the river becomes the border between Zambia and the eastern Caprivi Strip of Namibia for about 130 km.

The Chobe tributary originates in Angola, crosses the Caprivi Strip with an annual discharge of about 1.3 km^3, then forms the border between Namibia and Bostwana and enters Botswana to flow southwards for about 75 km until it meets the Selinda spillway along which spillage from the Okavango occurs in high flood years (see section *The Okavango basin*). It then turns east, again forming the border between Namibia and Botswana as it flows through a swampy area and flows into the Zambezi River at the border point between Namibia, Botswana, Zimbabwe and Zambia with an annual discharge of about 4.1 km^3. The discharge of the Zambezi River at this point is 33.5 km^3/year.

The Zambezi River then forms the border between Zambia and Zimbabwe and reaches its greatest width, over 1.3 km, before its waters plunge over the Victoria Falls. It continues to form the border between Zambia and Zimbabwe until it enters Mozambique.

There are two major man-made lakes on the Zambezi River, Lake Kariba on the border between Zambia and Zimbabwe and Lake Cabora Bassa in Mozambique.

Downstream of Lake Kariba the Kafue River, a major tributary originating in the north of Zambia, flows into the Zambezi River with a discharge of about 10 km^3/year. Still further downstream, at the border with Mozambique, the Luangwa River flows into the Zambezi River with an annual discharge of over 22 km^3. This tributary originates in the north-east of Zambia. The total discharge entering Lake Cabora Bassa from Zambia is estimated at about 77.5 km^3/year.

Leaving the lake the Zambezi River flows south-eastwards and receives water from its last great tributary, the Shire, with an annual discharge of nearly 16 km^3. The Shire drains Lake Malawi (also called Lake Nyasa) about 450 km to the north. The northern part of Lake Malawi forms the border between Tanzania and Malawi, the southern part the border between Mozambique and Malawi. The total flow into the lake is estimated at about 29 km^3/year of which 53% from Tanzania, 43% from Malawi and 4% from Mozambique. Total outflow from the lake in the Shire River in the south is 12.5 km^3/year. The level of the lake has fluctuated 6 metres since the beginning of the century, with its lowest level in 1917 and its highest level in 1980.

At its mouth, the Zambezi River splits into a wide, flat and marshy delta. The annual discharge flowing to the sea is estimated at 106 km^3.

Annual rainfall in the basin decreases from almost 1 800 mm in the north to less than 550 mm in the south. Both Botswana and Namibia are rather dry countries and only 2% of each of these countries is situated in the basin. However, rainfall in these parts, around 600 mm/year, is higher than the countries' average, which is 400 mm/year for Botswana and only 280 mm/year for Namibia.

Irrigation potential and water requirements

For **Angola** the irrigation potential has been estimated at 700 000 ha [*], as explained in the section *The Congo/Zaire basin*.

The irrigation potential for **Namibia** has been estimated at between 45 000 and 50 000 ha, of which 10 000 ha for a sugar cane project in the Caprivi Strip [163]. Flood recession cropping is evaluated at 1 000 ha in this area.

The irrigation potential for **Botswana** in the Zambezi basin ranges from 80 ha, considering identified soils and without the need for major water development works, to 11 080 ha, including the need for major water development works. However, of this total area, 10 000 ha are located in the Padamatenga plains outside the Zambesi basin in the north-east, to where it is planned to transfer water from the Chobe tributary [64]. In this study, 1 080 ha have been retained for the irrigation potential.

According to the irrigation subsector review of **Zimbabwe** [216], of the total internal surface water resources of 13.1 km^3/year, 3.6 km^3/year is already committed for domestic, industrial, mining and irrigation use. Of the remaining 9.5 km^3/year, at least 3.0 km^3/year is reported to be effectively inaccessible. Of the remaining 6.5 km^3/year, about half is considered as potentially available for irrigation development, of which 1.94 km^3/year in the Zambezi basin. In addition, there is the flow of the Zambezi River.

The Zambezi basin in Zimbabwe has been divided into three hydrological zones. In the western and eastern zones, suitable soil is the limiting factor, while in the middle zone water is the limiting factor. At present 70 045 ha have been developed or planned for irrigation [216]. Based on land and water and considering an irrigation water requirement of 10 500 m^3/ha per year according to this study, it would be possible to irrigate another 95 355 ha, so bringing the total to 165 400 ha [*]. However, taking a water requirement of 18 000 m^3/ha per year as proposed in the irrigation subsector review would reduce the potential to 131 000 ha.

For **Zambia**, of the irrigation potential of 523 000 ha for the whole country the distribution of 355 000 ha over the different sub-basins is known, but no details on location are given for the remaining 168 000 ha, consisting of 100 000 ha of dambos (wetlands), 60 000 ha irrigated by renewable groundwater and 8 000 ha of commercial farms [214, 215]. The irrigation potential in the Zambezi basin is estimated at 422 000 ha as follows [215, *]:

TABLE 39
Irrigation potential in the different Zambezi sub-basins in Zambia

Types of irrigation	Upper Zambezi River basin (ha)	Kafue River basin (ha)	Luangwa River basin (ha)	Total for Zambezi River basin (ha)
Located	112 000	165 000	14 000	291 000
Groundwater	15 000	15 000	15 000	45 000
Commercial	2 000	2 000	2 000	6 000
Dambos	30 000	20 000	30 000	80 000
Total	159 000	202 000	61 000	422 000

In view of the rugged, very steep terrain of the southern highlands of **Tanzania** draining to Lake Malawi, no real possibilities for irrigation development exist and consumptive water use would be limited to domestic and industrial water supply. These are relatively small volumes and their quantities would not change the mean annual flow into the lake from Tanzania [136].

Malawi has abundant land where soil and topography are suitable for irrigation but only limited areas that coincide with easily obtainable water from perennial streams. An important feature of the flat lake shore and the Shire River valley landscapes are large areas of marshy land, swamps and lagoons, which are poorly drained and flood susceptible areas. The irrigation potential for the whole of Malawi has been estimated at 100 000 ha plus 61 900 ha of dambos (wetlands) [135]. It is estimated that 160 900 ha of this total are located in the Zambezi basin [*].

The irrigation pðtential figure of 2 million ha given in literature for **Mozambique** includes the whole Zambezia province, part of which is located outside the Zambezi basin [159]. The area within the Zambezi basin is estimated at 1.7 million ha [*].

Table 40 summarizes the figures for the whole basin.

TABLE 40
Zambezi basin: irrigation potential, water requirements and areas under irrigation

Country	Irrigation potential (ha)	Gross potential irrigation water requirement		Area under irrigation (ha)
		per ha (m^3/ha per year)	total (km^3/year)	
Angola	700 000	13 500	9.450	2 000
Namibia	11 000	5 000 - 25 000	0.255	6 142
Botswana	1 080	5 500	0.006	0
Zimbabwe	165 400	10 500	1.737	49 327
Zambia	422 000	12 000	5.064	41 400
Tanzania	0	11 000	0.000	0
Malawi	160 900	13 000	2.092	28 000
Mozambique	1 700 000	11 000	18.700	20 000
Sum of countries	3 160 380		37.303	146 869
Total for Zambezi	3 160 380		37.303	

For the Zambezi basin as a whole, the water requirements are much less than the water availability. Attention has to be paid, however, to the Chobe tributary, originating in Angola and shared by Angola, Zambia, Namibia and Botswana. The Zambezi River entering Zambia from Angola in the north has an annual discharge of 18 km^3, which is twice the volume needed to irrigate the 700 000 ha potential of Angola. The Chobe tributary, however, has a discharge of only 1.3 km^3/year when leaving Angola, so if a large part of the irrigation potential area of 700 000 ha in Angola or if part of the irrigation potential of 159 000 ha in the upper Zambezi basin in Zambia is located in this sub-basin, problems may arise for Namibia and Botswana, even though irrigation potential there is very limited compared to the other countries.

Further downstream, no particular problems are expected in terms of water resources. However, water regulation would be necessary for full development of the potential.

The Okavango basin

The Okavango basin covers 1% of the continent. It is an endorheic basin, shared between Angola, Namibia and Botswana (Map 9 and Table 41).

TABLE 41
Okavango basin: areas and rainfall by country

Country	Total area of the country (km^2)	Area of the country within the basin (km^2)	As % of total area of basin (%)	As % of total area of country (%)	Average annual rainfall in the basin area (mm)		
					min.	max.	mean
Angola	1 246 700	166 963	51.7	13.4	525	1 320	865
Namibia	824 900	106 798	33.0	12.9	355	595	465
Botswana	581 730	49 431	15.3	8.5	415	570	495
For Okavango		323 192	100.0		355	1 320	680

Rivers and discharges

The two main rivers, the Cubango and Cuito, originate in Angola and flow to the south, where they become the border between Angola and Namibia. After flowing together they become the Okavango River that enters the Caprivi Strip in Namibia about 50 km further downstream. The average annual discharge leaving Angola at Mukwe is 10 km^3.

The Omatako tributary in Namibia is an ephemeral river, flowing north-east to enter the Cubango River at the border between Angola and Namibia.

After entering Botswana, the Okavango River flows into the Okavango Delta, a large swamp area. A spillway exists from this area to the Chobe River in the Zambezi basin in periods of high floods.

Irrigation potential and water requirements

The ecological value of the Okavango region is high and increasing abstractions of water for irrigation purposes may have a negative effect on the ecology of the Caprivi Strip area in Namibia and the Okavanga Delta in Botswana [163]. This requires a very judicious use of the water resources by the three riparian countries.

The irrigation potential in **Angola** has been estimated at 200 000 ha, as explained in the section *The Congo/Zaire basin* [*].

The irrigation potential of **Namibia** has been estimated at 2 000 ha, of which 1 000 ha for flood recession cropping [*]. The eastern national water carrier project in Namibia plans to transfer 60 million m^3/year of water from the Okavango River to the central and western coastal areas of the country. However, it is planned to use this water mainly for domestic purposes and not for agriculture [162].

The maximum irrigation potential in the Okavango region in **Botswana** has been estimated at about 9 060 ha, of which 3 000 ha would need important constructions for water development and storage [64]. In this study, the figure of 6 060 ha is retained for the irrigation potential [*].

TABLE 42
Okavango basin: irrigation potential, water requirements and areas under irrigation

Country	Irrigation potential (ha)	Gross potential irrigation water requirements		Area under irrigation (ha)
		per ha (m^3/ha per year)	total (km^3/year)	
Angola	200 000	7 000	1.400	0
Namibia	2 000	5 500	0.011	0
Botswana	6 060	6 000	0.036	0
Sum of countries	208 060		1.447	0
Total for Okavango	208 060		1.447	

The Limpopo basin

The Limpopo basin, located in South-eastern Africa, covers 1.3% of the continent and spreads over four countries (Map 10 and Table 43).

TABLE 43
Limpopo basin: areas and rainfall by country

Country	Total area of the country (km^2)	Area of the country within the basin (km^2)	As % of total area of the basin (%)	As % of total area of the country (%)	Average annual rainfall in the basin area (mm)		
					min.	max.	mean
Botswana	581 730	80 118	19.9	13.8	290	555	425
Zimbabwe	390 760	51 467	12.8	13.2	300	635	465
South Africa	1 221 040	185 298	46.1	15.2	290	1 040	590
Mozambique	801 590	84 981	21.1	10.6	355	865	535
For Limpopo		401 864	100.0		290	1 040	530

Rivers and discharges

The Crocodile River, which is the upper part of the Limpopo River, originates in South Africa near Johannesburg. It flows north-westwards to the border with Botswana and then turns to flow north-eastwards, first on the border between South Africa and Botswana and then on the border between South Africa and Zimbabwe. Several tributaries originate in Botswana, the most important being the Shashi, which forms the border between Botswana and Zimbabwe before flowing into the Limpopo River. Entering Mozambique, the river has an average annual discharge of 4.8 km^3.

Another important tributary, the Elephants River (also called the Transvaal River), originates in South Africa not far from Johannesburg and flows in north-eastwards. It flows into the Limpopo River in Mozambique.

The Mozambican part of the basin area is estimated to contribute only 10% of the total mean annual runoff of the river [155]. The Limpopo River, which was initially a perennial river in Mozambique, can actually fall dry for up to a period of eight months per year, mainly due to abstractions in the upper catchment area [155].

Irrigation potential and water requirements

The quantity of water produced in the Limpopo basin within **Botswana** is estimated at about 0.6 km^3/year [61]. The maximum irrigation potential is estimated at 15 208 ha, of which about 10 000 ha would need important works for water development and storage [64]. Moreover, as several major towns of Botswana are located in this area, including the capital Gabarone at the Notwane tributary, water problems may arise. This study has retained an irrigation potential of 5 000 ha for this region [*].

Surface water resources produced in the basin area in **Zimbabwe** are estimated at 0.54 km^3/year, of which 0.41 km^3 drains to the Limpopo River at the border between Zimbabwe and South Africa and 0.13 km^3 enters Mozambique before flowing into the Limpopo River. After deducting the water already committed for domestic, industrial, mining and irrigation use and the water which can not be developed, about 0.076 km^3/year of water is considered as being potentially available for irrigation development. At present 3 992 ha have been developed or planned for irrigation [216]. Land still suitable for irrigation is about 70 000 ha, but water constraints limit the area to 6 900 ha [*]. This brings the total irrigation potential to about 10 900 ha.

For **South Africa** the water resources per sub-basin are known [190]. It is estimated that by 2010 in the whole of South Africa 15 to 16 km^3/year of water will be available for agricultural purposes. Table 44 summarizes the irrigated areas, water availability, water requirements and irrigation potential for the Limpopo basin in South Africa [190, *].

At present 198 000 ha are already irrigated, using less than 10 000 m^3/ha per year instead of the 12 000 m^3/ha per year estimated in the present study.

The irrigation potential for **Mozambique** in the Limpopo basin has been estimated at 148 000 ha [159].

TABLE 44
Irrigated areas, water availability, water requirements and irrigation potential in the Limpopo basin in South Africa

Sub-basin	Irrigated area (ha)	Actual water use (km^3/yr)	Water available (km^3/yr)	Irrigation water requirement (m^3/ha.yr)	Irrigation potential (ha)
Crocodile (A)	95 000	1.090	0.813	12 000	67 800
Elephants (B)	103 000	0.768	0.765	12 000	63 700
Total for Limpopo basin in South Africa	198 000	1.858	1.578		131 500

Table 45 summarizes the irrigation potential, water requirements and irrigated areas for the Limpopo basin.

TABLE 45
Limpopo basin: irrigation potential, water requirements and area under irrigation

Country	Irrigation potential (ha)	Gross potential irrigation water requirements		Area under irrigation (ha)
		per ha (m^3/ha per year)	total (km^3/year)	
Botswana	5 000	10 500	0.053	1 381
Zimbabwe	10 900	11 000	0.120	2 000
South Africa	131 500	12 000	1.578	198 000
Mozambique	148 000	11 500	1.702	40 000
Sum of countries	295 400		3.452	241 381
Total for Limpopo	$\leq$ 295 400		3.452	

In view of the fact that the Limpopo River in Mozambique can already fall dry during eight months of the year, the above potential has to be considered as an upper limit, requiring important storage works and good cooperation between the basin countries.

Should South Africa use 12 000 m^3/ha per year, it would already irrigate more than its potential. For Botswana and Zimbabwe the literature gives higher irrigation water requirements than the present study, which means that in such a case the already small potential of these countries would also have to be reduced.

The Orange basin

The Orange basin, located in Southern Africa, covers almost 3% of the continent and spreads over four countries (Map 11 and Table 46).

TABLE 46
Orange basin: areas and rainfall by country

Country	Total area of the country (km^2)	Area of the country within the basin (km^2)	As % of total area of basin (%)	As % of total area of country (%)	Average annual rainfall in the basin area (mm)		
					min.	max.	mean
Botswana	581 730	71 000	7.9	12.2	165	520	295
Namibia	824 900	219 249	24.5	26.6	35	415	185
Lesotho	30 350	30 350	3.4	100.0	575	1 040	755
South Africa	1 221 040	575 769	64.2	47.2	35	1 035	365
For Orange basin		896 368	100.0		35	1 040	325

Rivers and discharges

The source of the Orange River is in Lesotho. The river receives water from the Makhaleng tributary just before entering South Africa. The Caledon tributary flows on the border between South Africa and the north of Lesotho and flows into the Orange River further downstream in

South Africa. The average annual runoff from Lesotho to South Africa is estimated at 4.73 km^3/year, which is far in excess of the country's water requirements.

Almost the entire plateau of South Africa, representing over 48% of the area of the country, is drained by the Orange River and its tributaries, though they contribute only about 22% of the total runoff of South Africa. The Vaal is the major tributary of the Orange River and the average annual runoff in the Vaal basin area is about 4.27 km^3, of which 2.15 km^3 is exploitable. The average annual runoff of the Orange basin, excluding the Vaal, is estimated at 7.59 km^3, of which 5.76 km^3 is exploitable.

The Molopo, which forms the border between Botswana and South Africa, is a fossil river, which once flowed into the Orange River. Now it receives most of its very occasional flows from its tributaries in the northern Cape province of South Africa [61].

The Orange River forms the border between the south of Namibia and South Africa. The most important tributary entering from Namibia is the Fish River, on which the Hardap dam was constructed in 1972.

Irrigation potential and water requirements

The irrigation potential in the orange basin in **Botswana** is negligible because of a lack of renewable water resources [64].

Namibia currently has access to an agreed volume of at least 0.5 km^3/year of water from the Orange River [162]. The gross irrigation water requirement is estimated at 8 500 m^3/ha per year in the present study. Literature gives figures of 33 000 m^3/ha per year [163]. The difference can be explained by a difference in the assumed future irrigation cropping pattern. As probably good soils alongside the rivers will be a limiting factor, an irrigation potential of 25 000 ha has been retained [*].

The soils of **Lesotho**, which lies entirely in the Orange basin, are very poor [192]. The irrigation potential has been estimated at 12 500 ha [127]. Through the Lesotho highlands water project, it is planned to transfer about 2.1 km^3/year of water from Lesotho to South Africa while enabling Lesotho to generate its own electricity.

The water available for agriculture in the Orange basin in **South Africa** is estimated at 4.3 km^3/year by the year 2010 [190]. Table 47 summarizes the water resources, irrigation potential and water requirements [190, 191, *].

TABLE 47
Irrigated areas, water availability, water requirements and irrigation potential in the Orange basin in South Africa

Sub-basin	Actual irrigated (ha)	Actual water use (km^3/yr)	Water available (km^3/yr)	Irrigation water requirement (m^3/ha.yr)	Irrigation potential (ha)
Vaal (C)	160 000	1.366	1.770	14 000	126 400
Orange (D)	140 000	1.413	2.488	11 000	226 100
Total for Orange basin in South Africa	300 000	2.779	4.258		352 500

For the whole of the Orange basin in South Africa, the available water of 4.258 km^3/yr would lead to an irrigation potential of 352 500 ha, using between 11 000 and 14 000 m^3/ha of water per year [*]. At the moment 300 000 ha already benefit from irrigation, using less than 10 000 m^3/ha per year.

Table 48 below summarizes the irrigation potential, water requirements and irrigated areas for the Orange basin.

TABLE 48
Orange basin: irrigation potential, water requirements and areas under irrigation

Country	Irrigation potential (ha)	Gross potential irrigation water requirement		Area under irrigation (ha)
		per ha (m^3/ha per year)	total (km^3/year)	
Botswana	0	7 000	0.000	0
Namibia	25 000	8 500 - 33 000	0.500	0
Lesotho	12 500	9 500	0.119	2 722
South Africa	352 500	11 000 - 14 000	4.257	300 000
Sum of countries	390 000		4.875	302 722
Total for Orange basin	≤ 390 000		4.875	

The irrigation potential of 390 000 ha for the Orange basin should be considered as an upper limit from the point of view of water availability. It would require agreed rules for the management of the river water by Namibia and South Africa.

The South Interior

The South Interior is divided into two separate basins, as shown in Figure 2. One is shared by Zimbabwe, Botswana and Namibia. A major part of the Kalahari Desert is located in this basin. The other one is shared by Angola and Namibia. Its total area represents 2.1% of the area of the continent (Map 12 and Table 49).

TABLE 49
South Interior: areas and rainfall by country

Country	Total area of the country (km^2)	Area of the country within the basin (km^2)	As % of total area of basin (%)	As % of total area of country (%)	Average annual rainfall in the basin area (mm)		
					min.	max.	mean
Zimbabwe	390 760	24 210	3.7	6.2	465	660	550
Botswana	581 730	368 780	57.1	63.4	270	670	405
Angola	1 246 700	53 118	8.2	4.3	500	905	680
Namibia	824 900	199 718	30.9	24.2	275	580	410
For South Interior		645 826	100.0		270	905	435

Rivers and discharges

The surface water resources of Zimbabwe are estimated at 0.038 km^3/year, of which 0.008 km^3 is still available for irrigation development after deducting quantities already used or committed [216]. The annual runoff of the Mosupe and Mosetse rivers, located in Botswana, is estimated at 0.055 km^3. Most of the rivers are ephemeral. In Angola the South Interior occupies 4% of the area of the country, but no information is available on water resources. In the Namibian part of the basin there are only ephemeral rivers.

Irrigation potential and water requirements

The area already developed or planned for irrigation in **Zimbabwe** is about 250 ha [216]. The combination of suitable land, which is already limited, and available water, which is even more of a limiting factor, leads to an irrigation potential of 1 100 ha, considering an irrigation water requirement of 9 500 m^3/ha per year [*]. Should 18 000 m^3/ha per year of water be used, as planned in the irrigation subsector review of Zimbabwe [216], the area would have to be even less than 1 100 ha.

The maximum irrigation potential for **Botswana** in the South Interior has been estimated at 3 950 ha, all of it being located in the Makgadikgadi Pans in the eastern part of Botswana. About 1 450 ha are marginal land and would need major constructions for water development and storage [64]. In this study, 2 500 ha have been retained for the irrigation potential. The Central Kalahari Game Reserve in Botswana occupies a large area of the South Interior.

For **Angola**, no details are available on the distribution of the irrigation potential over the country. It has been estimated that 50 000 ha are located in the South Interior (see the section *The Congo/Zaire basin*) [*].

In **Namibia**, in the part of the South Interior which is shared with Angola, a large area is occupied by the Etosha National Park. Irrigation potential has been estimated at 400 ha [*]. The present study estimates a water requirement of 5 500 m^3/ha per year. Other estimates, considering a different irrigation cropping pattern, go as high as 34 000 m^3/ha per year [163].

TABLE 50
South Interior: irrigation potential, water requirements and areas under irrigation

Country	Irrigation potential (ha)	Gross potential irrigation water requirement		Area under irrigation (ha)
		per ha (m^3/ha per year)	total (km^3/year)	
Zimbabwe	1 100	9 500	0.010	250
Botswana	2 500	6 000	0.015	0
Angola	50 000	5 000	0.250	0
Namibia	400	5 500 - 34 000	0.008	0
Sum of countries	54 000		0.283	250
Total for South Interior	≤ 54 000		0.283	

The North Interior

The North Interior, which corresponds to the Sahara Desert, occupies almost 20% of the African continent. It extends from Morocco in the west to Egypt in the east. The largest part is occupied by Algeria (33%) and Libya (25%). More than 80% of the area of each of these two countries is located in this region (Map 13 and Table 51). The average annual rainfall is only 40 mm. It is even 0 mm in Niger.

TABLE 51
North Interior: areas and rainfall by country

Country	Total area of the country (km^2)	Area of the country within the basin (km^2)	As % of total area of basin (%)	As % of total area of country (%)	Average annual rainfall in the basin area (mm)		
					min.	max.	mean
Morocco + W.Sahara	712 500	154 682	2.7	21.7	0	455	95
Mauritania	1 025 520	578 393	10.0	56.4	0	465	30
Mali	1 240 190	512 746	8.8	41.3	0	700	70
Algeria	2 381 740	1 944 795	33.5	81.7	0	520	45
Tunisia	163 610	78 448	1.4	47.9	17	345	90
Niger	1 267 000	11 316	0.2	0.9	0	0	0
Libya	1 759 540	1 472 372	25.4	83.7	0	390	10
Chad	1 284 000	217 465	3.7	16.9	0	305	45
Sudan	2 505 810	313 365	5.4	12.5	0	315	105
Egypt	1 001 450	520 881	9.0	52.0	0	100	15
For North Interior		5 804 463	100.0		0	700	40

Rivers and discharges

The renewable water resources in the Rheris and Guir basins in Morocco are estimated at 0.82 km^3/year, of which 0.67 km^3/year is surface water and 0.15 km^3/year groundwater. No information is available about the Western Sahara. Average rainfall is 30 mm/year in Mauritania and 70 mm/year in Mali. No information is available on renewable water resources in these countries.

For Algeria water availability and needs for 2025 have been estimated by basin [50]. Table 52 summarizes the figures for the five basins of the North Interior part of Algeria.

Available renewable water resources in the North Interior in Tunisia are estimated at 0.20 km^3/year, of which 0.15 km^3 is surface water and 0.05 km^3 groundwater. For the whole of Tunisia it is estimated at 2.8 km^3/year (of which 2.1 km^3 is surface water [206]), which is about 80% of the total internal renewable water resources, estimated at 3.52 km^3/year.

TABLE 52
Estimated water balance in the North Interior in Algeria in 2025 [50]

(in km^3/year)	2025
Water availability:	
Total surface water	1.060
Available surface water	0.158
Groundwater (1)	2.051
Water re-use	0.678
Total available water	2.887
Water demands:	
Irrigation	2.503
Other water uses	0.983
Total water use	3.486
Balance	- 0.599

(1) About 1.683 km^3 is considered to be fossil water

Rainfall in the North Interior in Niger, occupying less than 1% of the country, is negligible. Average annual rainfall is 10 mm in Libya, 45 mm in Chad, 105 mm in Sudan and 15 mm in Egypt. No information is available on renewable water resources in the North Interior for these countries.

Irrigation potential and water requirements

At present the irrigation water use of **Morocco** is 13 375 m^3/ha per year, in 2020 the country estimates its water use for irrigation to be 10 380 m^3/ha per year [149]. In the present study the water requirements for Morocco range from 7 000 m^3/ha per year in the north to 15 000 m^3/ha per year in the south. The irrigation potential has been estimated at 60 000 ha [152]. Should 15 000 m^3/ha per year of water be used in the North Interior, the annual water use would exceed the water available, and the area should be reduced to 40 000 ha at most [*].

The irrigation potential for **Mauritania** and **Mali** using renewable water resources has been considered negligible [*].

In **Algeria** the irrigation water requirement has been estimated at 7 000 m^3/ha per year [49]. The present study considers an irrigation water requirement of 12 000 m^3/ha per year for the northern part of the basin, where irrigation schemes would possibly be situated. The irrigation potential using renewable water ranges from a minimum of 20 000 ha, considering a water use of 12 000 m^3/ha per year and a water availability of 0.240 km^3/year, to a maximum of 125 200 ha, considering a water use of 7 000 m^3/ha per year and a water availability of 0.876 km^3/year. The low estimate on water availability is based on the assumption that where the total water balance is negative this quantity is deducted from the water available for irrigation [*]. The planned use of 1.627 km^3/year of fossil water for irrigation in 2025 would lead to irrigation of 135 600 if using 12 000 m^3/ha per year of water and 232 450 ha if using 7 000 m^3/ha per year of water [50, *].

The irrigation potential for the whole of **Tunisia** has been estimated at 563 000 ha, of which 40 000 ha in the North Interior [204]. With an irrigation water requirement of 14 500 m^3/ha per year of water [*], the total water requirement would be 0.58 km^3/year, which greatly exceeds the total available water resources, estimated at 0.20 km^3/year. Reducing the area to 11 000 ha means 0.16 km^3/year of water would be required. The 40 000 ha could only be irrigated by a water use of 4 000 m^3/ha per year.

As there is no rainfall in the North Interior part located in **Niger**, there is no potential for irrigation from renewable water resources [*].

The renewable water resources of the whole of **Libya** are estimated at 0.6 km^3/year. As at present the agricultural water consumption is already 4.275 km^3/year, most of it is fossil water [131]. It is planned to use an additional 2.365 km^3/year by 2025, bringing the total to 6.640 km^3/year. The irrigable land has been evaluated at 750 000 ha [131]. Estimating 250 000 ha to be located in the North Interior [*] would require 2.225 km^3/year of water if using of 8 900 m^3/ha per year of water [131]. Taking an irrigation water requirement of 18 000 m^3/ha per year [*] would lead to a total irrigation water requirement of 4.500 km^3/year, all fossil water.

The oases in **Chad** are estimated to cover in total 100 000 ha [85]. It is not clear what water would be used to irrigate these 100 000 ha, but probably most will be fossil water. This study estimates 10 000 ha to be located in the North Interior. The area in **Sudan** is also estimated at 10 000 ha, in **Egypt** 50 000 ha, all dependent mainly on fossil water [*].

TABLE 53

North Interior: irrigation potential, water requirements and areas under irrigation

Country with an area within the North Interior	Irrigationl potential using renewable water (ha)	Gross irrigation water requirement (m^3/ha.year)	Total irrigation water req. (km^3/year)	Irrigation potential using fossil water (ha)	Gross irrigation water requirement (m^3/ha.year)	Total irrigation water req. (km^3/year)	Area already under irrigation (ha)
Moroc + W.Sah (1)	60 000	10 000-15 000	0.600				10 000
Mauritania	0			2 000	17 500	0.035	2 000
Mali	0			5 000	19 500	0.098	0
Algeria (2)	125 200	7 000-12 000	0.876	135 600-232 450	7 000-12 000	1.627	45 000
Tunisia (3)	40 000	4 000-14 500	0.160				25 000
Niger	0						0
Libya	0			250 000	8 900-18 000	2.225-4.500	150 000
Chad	0			10 000	21 000	0.210	0
Sudan	0			10 000	20 000	0.200	500
Egypt	0			50 000	17 500	0.875	0
Sum of countries	225 200		1.636	461 600-559 450		4.979-7.254	232 500
Total North Interior	≤ 71 000						

(1) Considering 0.600 km^3/year of renewable water available for irrigation:
irrigation potential 60 000 ha if using 10 000 m^3/ha per year and 40 000 ha if using 15 000 m^3/ha per year.

(2) Considering 0.876 km^3/year of renewable water available for irrigation and using 7 000 m^3/ha year:
irrigation potential 125 200 ha.
Considering 0.240 km^3/year of renewable water available for irrigation and using 12 000 m^3/ha per year:
irrigation potential 20 000 ha.

(3) Available renewable water 0.160 km^3/year:
irrigation potential 40 000 if using 4 000 m^3/ha.yr and 11 000 ha if using 14 500 m^3/ha per year.

71 000 ha is the sum of: 40 000 ha (Morocco) + 20 000 ha (Algeria) + 11 000 ha (Tunisia).

Depending on the irrigation water requirements, estimates of the irrigation potential in the North Interior, based on renewable water resources, range from 71 000 ha to 225 200 ha.

The Mediterranean Coast

The Mediterranean Coast extends from Morocco in the west to Egypt in the east and is the aggregation of a large quantity of small, independant coastal basins draining to the sea. Its total area represents 2.2% of the area of the continent and spreads over five countries (Map 14 and Table 54).

TABLE 54
Mediterranean Coast: areas and rainfall by country

Country	Total area of the country (km^2)	Area of the country within the basin (km^2)	As % of total area of basin (%)	As % of total area of country (%)	Average annual rainfall in the basin area (mm)		
					min.	max.	mean
Morocco	446 500	108 300	15.9	24.3	185	740	350
Algeria	2 381 740	133 327	19.6	5.6	270	895	495
Tunisia	163 610	85 162	12.5	52.1	60	735	300
Libya	1 759 540	287 168	42.3	16.3	5	430	90
Egypt	1 001 450	65 568	9.6	6.5	60	140	100
For Mediter.Coast		679 525	100.0		5	895	235

Rivers and discharges

The total renewable water resources for the different basins and regions in the Mediterranean Coast in Morocco are summarized in Table 55.

For Algeria a study has been done on the water availability and needs for 2025 by basin, as explained in the section *The North Interior*. Table 56 summarizes the figures for the eight basins of the Mediterranean coastal part of Algeria [50].

The Medjerda River in Tunisia is the country's major perennial stream. Flows fluctuate greatly with quantities in June and July amounting to less than one-twelfth of those in February. The available renewable water resources in the Mediterranean Coast in Tunisia are estimated at about 2.60 km^3/year, of which 1.95 km^3 is surface water and 0.65 km^3 is groundwater (see also the section *The North Interior*).

TABLE 55
Renewable water resources by basin of the Mediterranean Coast in Morocco

Basin/region	Renewable surface water (km^3/year)	Renewable groundwater (km^3/year)	Total renewable water (km^3/year)
Moulouya	1.30	0.70	2.00
Loukkos	1.60	0.03	1.63
Other	2.85	0.40	3.25
Total	5.75	1.13	6.88

TABLE 56
Estimated water balance in the Mediterranean Coast in Algeria in 2025 [50]

(in km^3/year)	2025
Water availability:	
Total surface water	12.050
Available surface water	4.454
Groundwater	1.391
Water re-use	1.616
Total available water	7.461
Water demands:	
Irrigation	2.695
Other water uses	3.691
Total water use	6.386
Balance	1.075

The renewable water resources for Libya are estimated at 0.6 km^3/year. Information on the renewable water resouces of Egypt in this area is not available.

Irrigation potential and water requirements

The irrigation potential in the Mediterranean Coast in **Morocco** has been estimated at 380 000 ha [150, *]. The irrigation water requirement is about 9 000 m^3/ha per year [*]. The country estimates that its water use, at present 13 375 m^3/ha per year, will be 10 380 m^3/ha per year in 2020 [149].

In **Algeria**, the irrigation water requirement has been estimated at 7 000 m^3/ha per year [49]. The present study considers an irrigation water requirement of 9 000 m^3/ha per year for the Mediterranean Coast. The irrigation potential using renewable water ranges from a minimum of 243 150 ha, considering a water use of 9 000 m^3/ha per year and a water availability of 2.188 km^3/year, to a maximum of 385 100 ha, considering a water use of 7 000 m^3/ha per year and a water availability of 2.695 km^3/year. The low estimate on water availability is based on the assumption that for the basins where the water balance is negative (three out eight with a total deficit of 0.507 km^3/year) this quantity is deducted from the water available for irrigation [*].

The irrigation potential for the whole of **Tunisia** has been estimated at 563 000 ha, of which about 523 000 ha are in the Mediterranean Coast [204]. With an irrigation water requirement of 11 000 m^3/ha per year [*], this would require in total 5.75 km^3/year of water, which greatly exceeds the available water resources, estimated at 2.60 km^3/year. A reduced area of 189 000 ha would require 2.08 km^3/year of water. The 523 000 ha could only be irrigated using 4 000 m^3/ha per year of water.

In **Libya** about 40 000 ha could be irrigated with renewable water [*], the remaining part of the potential, estimated at 460 000 ha in the Mediterranean Coast would have to be irrigated by fossil water [131]. This is part of the Great Man Made River Project, where fossil water is transferred from the North Interior to the Mediterranean Coast.

The irrigation potential in **Egypt** has been estimated at 60 000 ha, almost all using fossil water [*].

TABLE 57

Mediterranean Coast: irrigation potential, water requirements and areas under irrigation

Country with an area within the North Interior	Irrigation potential using renewable water (ha)	Gross irrigation water requirement (m^3/ha.year)	Total irrigation water req. (km^3/year)	Irrigation potential using fossil water (ha)	Gross irrigation water requirement (m^3/ha.year)	Total irrigation water req. (km^3/year)	Area already under irrigation (ha)
Morocco	380 000	9 000	3.420				248 200
Algeria (1)	385 100	7 000-9 000	2.696				510 500
Tunisia (2)	523 000	4 000-11 000	2.080				360 000
Libya	40 000	12 500	0.500	460 000	8 900	4.094	320 000
Egypt	0			60 000	13 000	0.780	168 000
Sum of countries For Mediter. Coast	1 328 100 ≤ 850 000		8.696	520 000		4.874	1 606 700

(1) Considering 2.695 km^3/year of water available for irrigation and using 7 000 m^3/ha per year: irrigation potential 385 100 ha.
Considering 2.188 km^3/year of water available for irrigation and using 9 000 m^3/ha per year: irrigation potential 240 000 ha.

(2) Available renewable water: 2.080 km^3/year:
irrigation potential 523 000 ha if using 4 000 m^3/ha per year and 190 000 ha if using 11 000 m^3/ha per year.

850 000 ha is the sum of: 380 000 (Morocco) + 240 000 (Algeria) + 190 000 (Tunisia) + 40 000 (Libya).

Depending on the irrigation water requirements, estimates of the irrigation potential in the Mediterranean Coast, based on renewable water resources, range from 850 000 ha to 1 291 100 ha.

The North West Coast

The North West Coast covers 2.2% of the continent and spreads over three countries (Map 15 and Table 58).

TABLE 58
North West Coast: areas and rainfall by country

Country	Total area of the country (km^2)	Area of the country within the basin (km^2)	As % of total area of the basin (%)	As % of total area of the country (%)	Average annual rainfall in the basin area (mm)		
					min.	max.	mean
Morocco+W.Sah.	712 500	449 518	67.0	63.1	6	680	150
Mauritania	1 025 520	204 385	30.5	19.9	20	310	95
Algeria	2 381 740	16 718	2.5	0.7	0	110	60
For N.West Coast		670 621	100.0		0	680	145

Rivers and discharges

The total renewable water resources for the different basins and regions in the North West Coast in Morocco are summarized in Table 59.

No information on renewable water resources is available for the Western Sahara, Mauritania OR Algeria.

TABLE 59
Renewable water resources by basin of the North West Coast in Morocco

Basin/region	Renewable surface water (km^3/year)	Renewable groundwater (km^3/year)	Total renewable water (km^3/year)
Sebou	6.60	2.90	9.50
Oum er Rbia	4.50	1.50	6.00
Souss-Massa	0.48	0.29	0.77
Draa	0.77	0.10	0.87
Other	3.59	1.43	5.02
Total	15.94	6.22	22.16

Irrigation potential and water requirements

The irrigation potential in the North West Coast in **Morocco** has been estimated at 1 200 000 ha [15]. The present study estimates the gross irrigation water requirement at 9 000 m^3/ha per year in the northern part and 15 000 m^3/ha per year in the southern part. The country estimates that its water use, at present 13 375 m^3/ha per year, will be 10 380 m^3/ha per year in 2020. No data are available on the Western Sahara.

TABLE 60
North West Coast: irrigation potential, water requirements and areas under irrigation

Country	Irrigation potential (ha)	Gross potential irrigation water requirement		Area under irrigation (ha)
		per ha (m^3/ha per year)	total (km^3/year)	
Morocco + W.Sahara	1 200 000	9 000 - 15 000	12.000	1 000 000
Mauritania	0	17 500	0.000	750
Algeria	0	14 500	0.000	0
Sum of countries	1 200 000		12.000	1 000 750
Total for N. West Coast	≤ 1 200 000		12.000	

For **Mauritania** the only potential is some 750 ha of oases [145]. Less than 1% of the total area of **Algeria** is located in the North West Coast. The irrigation potential in this area is estimated to be negligible.

The West Coast

The West Coast is the region grouping all the basins draining to the sea from Senegal to Nigeria. It covers 4.7% of the continent and spreads over 13 countries (Map 16 and Table 61).

TABLE 61
West Coast: areas and rainfall by country

Country	Total area of the country (km^2)	Area of the country within the basin (km^2)	As % of total area of basin (%)	As % of total area of country (%)	Average annual rainfall in the basin area (mm)		
					min.	max.	mean
Senegal	196 720	124 854	8.7	63.5	350	1 630	870
Gambia	11 300	11 300	0.8	100.0	800	1 115	955
Guinea Bissau	36 120	36 120	2.5	100.0	1 260	2 440	1 700
Guinea	245 857	119 502	8.4	48.6	1 300	3 080	2 085
Sierra Leone	71 740	71 740	5.0	100.0	1 870	3 395	2 690
Liberia	97 750	97 750	6.8	100.0	1 770	3 300	2 370
Mali	1 240 190	9 496	0.7	0.8	545	1 365	675
Burkina Faso	274 000	197 379	13.8	72.0	555	1 310	920
Côte d'Ivoire	322 462	298 692	20.9	92.6	1 050	2 310	1 370
Ghana	238 540	238 540	16.7	100.0	855	1 785	1 265
Togo	56 785	56 785	4.0	100.0	925	1 550	1 215
Benin	112 620	66 236	4.6	58.8	915	1 345	1145
Nigeria	923 770	101 802	7.1	11.0	1 090	2 595	1 505
For West Coast		1 430 196	100.0		350	3 395	1 435

In this section two international basins in this region have been treated sepArately, the Gambia River basin and the Volta basin. The other basins have been regrouped and called 'the West Coast, excluding the Gambia River and Volta basins'.

The Gambia River Basin

The Gambia River basin occupies about 5.4% of the West Coast and is shared among three countries (Map 16 and Table 62).

TABLE 62
Gambia basin: areas by country

Country	Total area of the country (km^2)	Area of the country within the basin (km^2)	As % of total area of the basin (%)	As % of total area of the country (%)
Guinea	245 857	8 000	10.3	3.3
Senegal	196 720	58 550	75.2	29.8
Gambia	11 300	11 300	14.5	100.0
For Gambia basin		77 850	100.0	

Rivers and discharges

The Gambia River has its sources in the high rainfall mountainous Fouta Djallon in the north of the Central Guinea region. The total quantity of water leaving Guinea for Senegal is estimated at 3 km^3/year.

The river then flows northwards to enter The Gambia in the extreme east of the country. Contradictory information exists about the discharges entering The Gambia. According to different sources, they range from 4 km^3/year [181, average of 1951-1990] to nearly 10 km^3/year [25]. Its flow is highly seasonal: the peak discharge is about 2 000 m^3/s, but for six months the inflow at the Gambian border is less than 10 m^3/s. In May it falls below 0.5 m^3/s.

Because of the flat topography of The Gambia and the low river discharges during the dry season, salt water moves up to about 70 km upstream in the wet season and 250 km upstream in the dry season. The tidal variation at the mouth is about 1.6 m [48a].

Irrigation potential and water requirements

The higher, upstream part of the basin in **Guinea** is badly eroded. Irrigation would be possible in the downstream part, where the potential has been estimated at 20 000 ha [*].

There are 60 000 ha of suitable soils in the Gambia basin in **Senegal** [48a]. It is planned to construct a dam at Kekreti for hydropower and this could irrigate an estimated 15 000 ha in Senegal and 55 000 ha in The Gambia [181].

Soils suitable for irrigation in **The Gambia** are estimated at 80 000 ha [48a]. There are about 104 200 ha of swamps, of which 33 500 ha are cultivated. Mangroves account for an additional 67 000 ha [111]. In the dry season, the salt tongue moves upstream at a rate of 15-20 km/month. It is thought that an additional withdrawal of 1 m^3/s would increase the penetration of the salt tongue by 1 km/month. The safe limit for irrigation from the Gambia River without major dam construction is, therefore, estimated to be no more than 2 400 ha in the dry season [48a]. However, if the planned Kekreti dam on the Gambia River in Senegal is constructed, it is expected that 15 000 ha can be irrigated in Senegal and 55 000 ha in the Gambia [181]. Moreover, this dam could contain salt intrusion during the dry season. The development of these 55 000 ha would require 0.275 km^3/year of water. A further 25 000 ha of mangrove cultivation would require 0.125 km^3/year of water.

TABLE 63
Gambia River basin: irrigation potential and water requirements

Country	Irrigation potential (ha)	Gross potential irrigation water requirement	
		per ha (m^3/ha per year)	total (km^3/year)
Guinea	20 000	16 000	0.320
Senegal	15 000	7 000	0.105
Gambia	80 000	5 000	0.400
Sum of countries	115 000		0.825
Total for Gambia basin	115 000		0.825

Although the annual irrigation water requirement is only 10% of the discharge, any water abstraction within the basin in the dry season should be studied very carefully until the Kekreti dam is constructed, in view of the low discharges in the dry season and the danger of increasing salt intrusion from the sea.

The Volta Basin

The Volta basin occupies almost 28% of the total West Coast and is shared between six countries (Map 16 and Table 64).

Rivers and discharges

The most upstream part of the Volta basin is located in Mali, where it occupies less than 1% of the area of the country. One river, the Sourou, crosses the border from Mali to Burkina Faso, but there is almost no flow in this river.

TABLE 64
Volta basin: areas by country

Country	Total area of the country (km^2)	Area of the country within the basin (km^2)	As % of total area of the basin (%)	As % of total area of the country (%)
Mali	1 240 190	9 496	2.4	0.8
Burkina Faso	274 000	183 000	46.4	66.8
Benin	112 620	16 000	4.1	14.2
Togo	56 785	26 700	6.8	47.0
Côte d'Ivoire	322 462	7 000	1.8	2.2
Ghana	238 540	152 000	38.6	63.7
For Volta basin		394 196	100.0	

Two-thirds of Burkina Faso are within the Volta basin. The Black Volta (Mouhoun), Red Volta (Nazinon) and White Volta (Nakambé) all have their sources in Burkina Faso.

The *Black Volta* originates in the south-west of the country, flows north-eastwards and then turns south. In the south, it becomes the border, first between Ghana and Burkina Faso and then between Ghana and Côte d'Ivoire. When leaving Burkina Faso, its discharge is about 5 km^3/year; when entering Ghana, it is about 6 km^3/year. The *Red Volta* originates in the central part of Burkina Faso, near Ouagadougou, and flows south-eastwards to the border with Ghana. After crossing the border, it joins the White Volta. The *White Volta* originates in the north of Burkina Faso and also flows south-eastwards to the border with Ghana. The total annual discharge leaving Burkina Faso through the Red and White Volta Rivers is estimated at 3.7 km^3/year.

The Pendjari River originates in the north-west of Benin. It flows north-east, then turns sharply to the west to become the border, first between Burkina Faso and Benin, then between Togo and Benin for just a short distance before entering Togo with a total annual discharge of 2.2 km^3. In Togo, which it crosses in the north, here called the Oti River. Further downstream it becomes the border between Togo and Ghana. Entering Ghana further south, its discharge is estimated at 11 km^3/year.

Many other tributaries have their source within Ghana, but especially in the northern savannah part most of these water courses run almost dry after the rains. The groundwater here is low yielding and cannot be relied upon for extensive irrigation [113]. In the south a dam has been constructed at Akosombo for hydropower. Behind this dam, one of the world's largest artificial lakes has been created, Lake Volta, with a surface area of 8 500 km^2 and a capacity of 148 km^3. The average annual discharge flowing to the sea is estimated at about 38 km^3.

Irrigation potential and water requirements

The irrigation potential in **Mali**, occupying less than 1% of the country and with very few surface water resources has been considered negligible [*].

The irrigation potential in **Burkina Faso** has been evaluated at 142 000 ha, distributed as follows over the different sub-basins [67, 69, 72, 73]:

Of these 142 000 ha, about 20 000 ha are valley bottoms and 7 000 ha small areas irrigated by small earth dams.

The irrigation potential of **Benin** has been evaluated at 300 000 ha, but no details are available on location [57]. It is estimated that 30 000 ha are located in the Pendjari Basin [*].

TABLE 65
Irrigation potential and water requirements by sub-basin in the Volta basin in Burkina Faso

Volta sub-basins	Irrigation potential (ha)	Water requirement (km^3/year)
Black Volta (Mouhoun-Sourou)	42 000	0.420
Bougouriba-Poni (tributaries of Black Volta)	30 000	0.300
Red Volta (Nazinon)	15 000	0.150
White Volta (Nakambé)	48 000	0.480
Ouglé (tributary of Oti)	7 000	0.070
Total	142 000	1.420

The irrigation potential of **Togo** has been evaluated at 180 000 ha, of which 100 000 ha are valley bottoms [21a]. No details are available on location. As the Volta basin occupies about half of Togo, half of the irrigation potential, or 90 000 ha, is estimated to be within the Volta basin [*].

Of the irrigation potential of 475 000 ha for the whole of **Côte d'Ivoire** [21a], 25 000 ha are estimated to be in the Volta basin [*].

The potential for irrigated rice production in the inland valley swamps and the floodplains within **Ghana** has been evaluated at 1.9 million hectares, of which 346 000 ha are estimated to be suitable for fully controlled irrigation development [114]. No figures are available on location. About two-thirds of the country being within the Volta basin, an irrigation potential of 1.2 million hectares has been tentatively estimated for this area [*].

TABLE 66
Volta basin: irrigation potential and water requirements

Country	Irrigation potential (ha)	Gross potential irrigation water requirement	
		per ha (m^3/ha per year)	total (km^3/year)
Mali	0	8 500	0.000
Burkina Faso	142 000	10 000	1.420
Benin	30 000	20 000	0.600
Togo	90 000	23 000	2.070
Côte d'Ivoire	25 000	20 000	0.500
Ghana	1 200 000	20 000	24.000
Sum of countries	1 487 000		28.590
Total for Volta basin	1 487 000		28.590

The total annual flow to the sea, 38 km^3, exceeds than the total annual irrigation water requirements for the whole basin, 28.5 km^3. Comparing the water requirements in the different parts of the basin with water availability, the balance remains positive everywhere.

The West Coast, excluding the Gambia River and Volta basins

Except for The Gambia, which is entirely located in the Gambia River basin, all the other countries from Senegal in the west to Nigeria in the East are partly or wholly located within this remaining part of the West Coast (Map 16 and Table 67).

Rivers and discharges

The area of Senegal in the West Coast can be divided into two parts:

- the area south of the Gambia basin: Casamance and Kayanga basins;
- the area north of the Gambia basin and south of the Senegal basin: Ferlo, Car-Car, Sine and Saloum basins.

TABLE 67
West Coast, excluding Gambia River and Volta basins: areas by country

Country	Total area of the country (km^2)	Area of the country within the basin (km^2)	As % of total area of the basin (%)	As % of total area of the country (%)
Senegal	196 720	66 304	6.9	33.7
Guinea Bissau	36 120	36 120	3.8	100.0
Guinea	245 857	111 502	11.6	45.4
Sierra Leone	71 740	71 740	7.5	100.0
Liberia	97 750	97 750	10.2	100.0
Burkina Faso	274 000	14 379	1.5	5.2
Côte d'Ivoire	322 462	291 692	30.4	90.5
Ghana	238 540	86 540	9.0	36.3
Togo	56 785	30 085	3.1	53.0
Benin	112 620	50 236	5.2	44.6
Nigeria	923 770	101 802	10.6	11.0
For West Coast, without Gambia and Volta basins		958 150	100.0	

The annual discharge of the Casamance River, as measured between 1968 and 1983 was 0.07 km^3. In the dry season (April-July) the river may run dry. Dams to protect the area against salt intrusion are necessary. The Kanyanga River is the upper part of the Gêba River in Guinea Bissau, but no discharge figures are available. Nor are there figures available for discharges in the northern part.

Guinea Bissau is wholly situated in the West Coast. The main rivers are the Cacheu originating within the country, the Gêba originating in Senegal and the Corubal originating in Guinea. The water resources in this small country are abundant, but they are badly distributed in space and in time: 90% of the flow occurs in 6 months. The annual discharge of the largest river, the Corubal, is estimated at 13.2 km^3/year. In the coastal area, problems of salt intrusion exist in the dry season and many 'anti-salt' dams have been constructed.

Two separate parts of Guinea are located in this West Coast area:

- the eastern part of the Middle Region and the Lower Region, draining to the sea;
- the southern part of the Forest Region, draining to Liberia and Sierra Leone.

The water resources of Guinea are abundant.

Sierra Leone is one of the most humid countries of Africa. It can be divided into 12 major river basins, of which five are shared with Guinea and two with Liberia.

Like Sierra Leone, Liberia is one of the most humid countries of Africa. Two types of river exist:

- the major basins from north-east to south-west, with rivers originating in Guinea and Côte d'Ivoire and with an average entering discharge of 15 to 20 km^3/year;
- numerous, short, coastal watercourses.

The source of the Comoé River is in the south-west of Burkina Faso, the most humid region of the country. It is one of the few permanent rivers of Burkina Faso, with an average annual discharge leaving the country to Côte d'Ivoire of about 1.29 km^3. In Côte d'Ivoire many other rivers run parallel southwards to the sea. In the west is the Cavally River, which has its source in Guinea, then enters Côte d'Ivoire and further downstream becomes the border between Côte d'Ivoire and Liberia.

In Ghana many rivers run more or less parallel southwards to the sea. The most important are the Pra, with an annual discharge of about 6.2 km^3, and the Tano, with 4.5 km^3.

The Mono originates in Togo and at about 100 km from the sea it becomes the border between Benin and Togo, with an average annual discharge of about 2.9 km^3. In the south-west of Togo is the Lake Togo basin, with an area of about 8 000 km^2. The Couffo originates at the border between Benin and Togo about 200 km north of the sea. In Benin, three main rivers flow southwards to the sea. The Ouémé originates in the centre. The Okpara tributary also originates in the centre but becomes the border between Nigeria and Benin before re-entering Benin to flow into the Ouémé. The discharge close to the sea is estimated at 5.4 km^3/year.

About one-third of the basin area of Nigeria is covered by tropical rain forest. Many rivers flow from north to south to the sea. The annual potential surface water resources of the basin area are estimated at 36 km^3. Peak outflows occur in September-October. Many dams have been built on the rivers of the western littoral, including the Oyan dam on the Oyan River. The runoff of the Osun River is regulated by the Asejire Dam.

Irrigation potential and water requirements

In **Senegal** about 22 000 ha in the Casamance are controlled region by small dams [179]. Considering a potential of 40 000 ha, the irrigation water requirement would be 0.920 km^3/year. This is much more than the average annual discharge measured in the river. Most of this area, however, consists of mangroves [*]. The potential for cereal production in the Kayanga region has been estimated at about 15 000 ha, which would lead to a water requirement of 0.105 km^3/year [179]. Although no detailed discharge figures exist, this is also more than the quantity available. In the part north of the Gambia basin the irrigation potential will probably not exceed 30 000 ha, based on available water resources [*]. In this region, it is planned to irrigate about 8 500 ha through the Cayor Canal, which in fact takes water from the Senegal River [181].

The irrigation potential in **Guinea Bissau** is estimated at 281 290 ha, which in fact corresponds to the total potential rice area, of which about 150 000 ha are mangroves [120].

The humid land potential in **Guinea** is estimated at 520 000 ha (140 000 ha of mangroves, 180 000 ha of wetland and 200 000 ha of floodplains), of which about 210 000 ha can be developed relatively easily for irrigated agriculture [116]. Table 68 below summarizes the different potential areas for the various regions and basins in Guinea.

TABLE 68
Humid land potential of Guinea by region and major basin group

Region	First category (ha)	Total of all humid area (ha) [*]	Distribution over river basin and coast (ha) [*]
Lower	60 886	180 000	West Coast: 180 000
Middle	22 204	80 000	Gambia basin: 20 000; West Coast: 60 000
Upper	109 224	190 000	Senegal basin: 5 000, Niger basin: 185 000
Forest	18 506	70 000	West Coast: 70 000
Total	210 820	520 000	

The irrigation potential for **Liberia** is estimated at 600 000 ha, consisting mainly of freshwater swamps [129]. The total area suitable for development in **Sierra Leone** has been estimated at 807 000 ha, ignoring environmental aspects of wetland development [185].

The irrigation potential in the Comoé basin in **Burkina Faso** is estimated at 17 460 ha [67, 69, 72, 73], of which about 2 500 ha are valley bottoms and 450 ha small areas irrigated by small earth dams. Of the irrigation potential of 475 000 ha for the whole of **Côte d'Ivoire**, 175 000 ha are valley, 200 000 ha are large floodplains and 100 000 ha are coastal swamps [21a]. Of this total, 400 000 ha are estimated to be in this part of the West Coast [*].

Of the irrigation potential of 300 000 ha for the whole of **Benin** [57], 170 000 ha are estimated to be located in this part of the West Coast [*]. For **Togo** it is estimated at 90 000 ha [*]. Of the total potential of 1.9 million hectares of **Ghana** [114], 700 000 ha are estimated to be located in this part of the West Coast [*].

In **Nigeria**, the national water resources master plan (NWRMP) estimates the tital identified irrigation potential in the basin area at 50 000 ha [172]. Irrigation projects in the region have not been accelerated compared to the other regions, because the hilly area would create some difficulties for the canal system. Water requirements are estimated at 16 500 m^3/ha per year in the present study and at 13 400 m^3/ha per year in the NWRMP. Water resources are abundant.

TABLE 69

West Coast, excluding Gambia River and Volta basins: irrigation potential and water requirements

Country	Irrigation potential (ha)	Gross potential irrigation water requirement	
		per ha (m^3/ha per year)	total (km^3/year)
Senegal	85 000	7 000 - 23 000	1.235
Guinea Bissau	281 290	23 000	6.470
Guinea	310 000	20 000	6.200
Sierra Leone	807 000	16 000	12.912
Liberia	600 000	16 000	9.600
Burkina Faso	17 460	24 000	0.419
Côte d'Ivoire	400 000	18 000	7.200
Ghana	700 000	16 000	11.200
Togo	90 000	18 000	1.620
Benin	170 000	18 500	3.145
Nigeria	50 000	13 400	0.670
Sum of countries	3 510 750		60.671
Total for remaining W.Coast	3 510 750		60.671

Most of the countries have abundant water resources and no water shortage problems will arise in irrigation development, except for Senegal and Guinea Bissau.

Table 70 summarizes the irrigation potential for the whole of the West Coast, including the Gambia River and Volta basins.

TABLE 70

West Coast: irrigation potential, water requirements and areas under irrigation

Country	Irrigation potential (ha)	Gross potential irrigation water requirement per ha (m^3/ha per year)	Gross potential irrigation water requirement total (km^3/year)	Area under irrigation (ha)
Senegal	100 000	7 000 - 12 000	1.340	0
Gambia	80 000	5 000	0.400	1 670
Guinea Bissau	281 290	23 000	6.470	17 115
Guinea	330 000	16 000 - 20 000	6.520	91 880
Sierra Leone	807 000	16 000	12.912	29 360
Liberia	600 000	16 000	9.600	2 100
Mali	0	8 500	0.000	0
Burkina Faso	159 460	10 000 - 24 000	1.839	23 480
Côte d'Ivoire	425 000	18 000 - 20 000	7.700	72 750
Ghana	1 900 000	16 000 - 20 000	35.200	6 374
Togo	180 000	18 000 - 23 000	3.690	7 008
Benin	200 000	18 500 - 20 000	3.745	9 146
Nigeria	50 000	13 400	0.670	50 000
Sum of countries	5 112 750		90.086	310 883
Total for West Coast	5 112 750		90.086	

The West Central Coast

The West Central Coast covers 2.3% of the continent and spreads over seven countries (Map 17 and Table 71).

TABLE 71

West central Coast: areas and rainfall by country

Country	Total area of the country (km^2)	Area of the country within the basin (km^2)	As % of total area of basin (%)	As % of total area of country (%)	Average annual rainfall in the basin area (mm) min.	max.	mean
Nigeria	923 770	58 493	8.3	6.3	1 420	2 740	2 070
Cameroon	475 440	239 021	33.9	50.3	1 365	2 830	1 835
Gabon	267 670	267 670	38.0	100.0	1 320	2 595	1 800
Equat. Guinea	28 050	28 050	4.0	100.0	1 695	2 585	2 050
Congo	342 000	95 023	13.5	27.8	1 125	1 940	1 475
Angola	1 246 700	7 150	1.0	0.6	775	1 280	1 110
Zaire	2 344 860	9 367	1.3	0.4	785	1 290	1 190
For West Central Coast		704 774	100.0		775	2 830	1 785

Rivers and discharges

Rising in the Cameroon highlands, an area of dense rain forest, the Cross river, enters Nigeria with an annual discharge estimated at 17 km^3. Annual runoff to the sea is estimated at almost 52 km^3. Another important river in Nigeria is the Imo River, with an average annual discharge of 4 km^3. The total surface water resources in the basin area are estimated at 69 km^3/year. About 85% of the annual runoff of the Cross River and 70% of the annual runoff of the Imo River are concentrated in five months, from June to October with the peak in September.

In Cameroon many rivers flow directly to the sea. The most important one is the Sanaga River, with an average annual discharge of almost 63 km^3. Other important rivers are the Nyong, the Wouri and the Ntem Rivers, with a total annual discharge of over 32 km^3.

Also in Gabon many rivers flow directly to the sea. The most important one is the Ogooué with an annual discharge of more than 148 km^3/year. Its basin occupies about 75% of the country. Another important river is the Nyanga to the south, with an annual discharge estimated at 17.1 km^3.

In the mainland part of Equatorial Guinea several watercourses, most of which originate within the country, cross the country while flowing to the sea. The renewable water resources are estimated at 30 km^3/year for the mainland and the island together.

Of the many rivers flowing to the sea in Congo, the most important one is the Kouilou-Niari River. Its basin covers nearly 60% of the area of Congo in the West Central Coast. Its annual flow to the sea is estimated at about 28 km^3.

Cabinda, the part of Angola lying in the West Central Coast, is separated from the rest of Angola by the Congo/Zaire River and a strip of land to the north of the river belonging to Zaire. Its area corresponds to only 0.6% of the total area of Angola. The most important river is the Chiloango, the upstream part of which forms the border between Zaire and Angola. The part of Zaire lying in the West Central Coast, only 0.4% of the total area of Zaire, corresponds to the basin of the Chiloango River.

Irrigation potential and water requirements

The identified irrigation potential in the West Central Coast in **Nigeria** is 100 000 ha according to the national water resources master plan (NWRMP) [172]. The irrigation water requirements are estimated at 15 000 m^3/ha per year in the present study and at 10 400 m^3/ha per year in the NWRMP. Water resources are abundant.

The irrigation potential in **Cameroon** has been estimated at 120 000 ha [*]. For **Gabon** it is 440 000 ha [17]. For **Equatorial Guinea** no figures on irrigation potential are available. It is estimated in the present study at 30 000 ha, corresponding to 1% of the total area of the country [*]. The irrigation potential figures of 85 000 ha for **Congo,** 50 000 ha for **Angola** and 10 000 ha for **Zaire** in the West Central Coast are explained in the section *The Congo/Zaire basin* [*].

TABLE 72
West Central Coast: irrigation potential, water requirements and areas under irrigation

Country	Irrigation potential (ha)	Gross potential irrigation water requirement		Area under irrigation (ha)
		per ha (m^3/ha per year)	total (km^3/year)	
Nigeria	100 000	10 400	1.040	20 000
Cameroon	120 000	15 000	1.800	3 000
Gabon	440 000	13 000	5.720	4 450
Equat. Guinea	30 000	12 500	0.375	
Congo	85 000	14 000	1.190	0
Angola	50 000	14 500	0.725	1 000
Zaire	10 000	14 500	0.145	0
Sum of countries	835 000		10.995	28 450
Total for W.Cent. Coast	835 000		10.995	

Many figures are arbitrary estimates as reliable information is lacking. It is a very humid region and neither land nor water is a limiting factor for the estimation of the irrigation potential.

The South West Coast

The South West Coast covers 1.7% of the continent and spreads over two countries (Map 18 and Table 73).

TABLE 73
South West Coast: areas and rainfall by country

Country	Total area of the country (km^2)	Area of the country within the basin (km^2)	As % of total area of basin (%)	As % of total area of country (%)	Average annual rainfall in the basin area (mm)		
					min.	max.	mean
Namibia	824 900	17 549	3.4	2.1	90	515	350
Angola	1 246 700	498 651	96.6	40.0	10	1 600	960
For S. West Coast		516 200	100.0		10	1 600	940

Rivers and discharges

Almost 97% of the area of the South West Coast is covered by Angola, the remaining part by Namibia, that shares the border river, the Cunene, with Angola. This river originates in the central highlands of Angola and its annual discharge reaching the border is about 5 km^3. Many other rivers originate within Angola. Annual rainfall in the South West Coast decreases considerably from the north-east to the south-west (from 1 600 mm to 10 mm).

Irrigation potential and water requirements

Namibia has access to an agreed volume of 0.18 km^3/year of water from the Cunene River, of which 0.13 km^3 is for agricultural purposes. Considering a gross irrigation water requirement of 5 500 m^3/ha per year [*], this would lead to an irrigation potential of 23 600 ha. However, literature gives estimates of gross irrigation water requirements of 16 500 m^3/ha per year [163], which would lead to an irrigation potential of 7 900 ha, using the 0.13 km^3/year of water.

The irrigation potential of **Angola** in the South West Coast is estimated at 1.8 million hectares, as explained in the section *The Congo/Zaire basin* [*].

TABLE 74
South West Coast: irrigation potential, water requirements and areas under irrigation

Country	Irrigation potential (ha)	Gross potential irrigation water requirement per ha (m^3/ha per year)	Gross potential irrigation water requirement total (km^3/year)	Area under irrigation (ha)
Namibia	7 900	16 500	0.130	0
Angola	1 800 000	13 500	24.300	70 000
Sum of countries	1 807 900		24.430	70 000
Total for S. West Coast	1 807 900		24.430	

The South Atlantic Coast

The South Atlantic Coast, located in South-Western Africa, covers 1.2% of the continent and spreads over two countries (Map 19 and Table 75).

TABLE 75
South Atlantic Coast: areas and rainfall by country

Country	Total area of the country (km^2)	Area of the country within the basin (km^2)	As % of total area of basin (%)	As % of total area of country (%)	Average annual rainfall in the basin area (mm)		
					min.	max.	mean
South Africa	1 221 040	101 325	27.7	8.3	45	555	200
Namibia	824 900	264 160	72.3	32.0	0	485	190
For S. Atl. Coast		365 485	100.0		0	555	190

Rivers and discharges

The South Atlantic Coast is the driest region in southern Africa. In Namibia a few ephemeral rivers exist, on which dams have been constructed. In South Africa three main basins are located in this region and the total surface water resources are estimated at 3.37 km^3/year, of which 1.62 km^3/year is exploitable and less than 1.00 km^3/year available for irrigation purposes [190].

Irrigation potential and water requirements

In **South Africa** the irrigation water requirements are estimated at 10 000 m^3/ha per year in this region [*]. Using 0.83 km^3/year of water for irrigation, this would lead to an irrigation potential of 83 200 ha. About 84 000 ha already benefit from irrigation, which means that from a point of view of water availability the maximum area has already been brought under irrigation.

The irrigation potential of **Namibia** in this region does not exceed 1 000 ha [*]. The present study estimates the water requirements at 7 500 m^3/ha per year, literature gives figures of 33 000 m^3/ha per year, as different irrigation cropping patterns are assumed.

TABLE 76
South Atlantic Coast: irrigation potential, water requirements and areas under irrigation

Country	Irrigation potential (ha)	Gross potential irrigation water requirement		Area under irrigation (ha)
		per ha (m^3/ha per year)	total (km^3/year)	
South Africa	83 200	10 000	0.832	84 000
Namibia	1 000	20 000	0.020	0
Sum of countries	84 200		0.852	84 000
Total for S. Atl. Coast	≤ 84 200		0.852	

The Indian Ocean Coast

The southern and south-western part of the Indian Ocean Coast is wholly situated in South Africa. The eastern part is shared between Swaziland, South Africa and Mozambique. The north-eastern part is shared between Zimbabwe and Mozambique (Map 20 and Table 77). Its total area represents 2.2% of the area of the continent.

TABLE 77
Indian Ocean Coast: areas and rainfall by country

Country	Total area of the country (km^2)	Area of the country within the basin (km^2)	As % of total area of basin (%)	As % of total area of country (%)	Average annual rainfall in the basin area (mm)		
					min.	max.	mean
Swaziland	17 364	17 364	2.6	100.0	600	1 020	780
South Africa	1 221 040	358 648	54.0	29.4	125	1 270	585
Zimbabwe	390 760	102 047	15.4	26.1	375	1 685	650
Mozambique	801 590	185 726	28.0	23.2	470	1 770	885
For Ind. Oc. Coast		663 785	100.0		125	1 770	680

Rivers and discharges

Of the four major rivers in Swaziland, two originate inside the country, the Mbuluzi and Ngwavuma rivers, and two in South Africa, the Komati and Usutu rivers. Total inflow from South Africa to Swaziland is 1.8 km^3/year. Total outflow from Swaziland is 3.5 km^3/year, of which 2.3 km^3 flow directly into Mozambique to the Umbulezi and the Maputo rivers. The remaining 1.2 km^3 first enter South Africa before flowing into Mozambique, in the south towards the Maputo River and in the north towards the Incomati River. The Sabie River is another tributary of the Incomati River originating in South Africa.

Within South Africa, perennial rivers occur over only one quarter of the area and mainly in the southern and south-western Cape province and on the eastern plateau slopes. However, even the perennial rivers are very irregular and have important seasonal variations. The surface water resources in the Indian Ocean part are estimated at 31 km^3/year, of which about 21 km^3/year are exploitable. Less than 10 km^3/year are available for agricultural purposes.

The Save, Buzi and Pungoé rivers originate in Zimbabwe and all flow to Mozambique. Although the catchment area of the Pungoé River in Zimbabwe is only 5% of the total catchment area, about 26% of the annual runoff originates from this area [155].

Irrigation potential and water requirements

For **Swaziland** the following irrigation potential figures are given for the country, based on land and water availability [197]:

TABLE 78

Water resources, irrigation potential and water requirements by sub-basin in Swaziland

Basin	Inflow from RSA (km^3/yr)	Produced in country (km^3/yr)	Outflow (km^3/yr)	Irrigation potential (ha)	Water demand (km^3/yr)
Lomati-Komati	0.738	0.415	1.153	17 925	0.161
Mbuluzi	0	0.352	0.352	24 280	0.219
Usutu	1.032	0.904	1.936	45 875	0.413
Ngwavuma	0	0.106	0.106	5 140	0.046
Total	1.770	1.777	3.547	93 220	0.839

In **South Africa**, about 9 km^3/year of water are estimated to be available for agricultural purposes in the Indian Ocean Coast in 2010 [190]. Table 79 summarizes the water resources, irrigation potential and water requirements for the different basins (see also Map 20).

TABLE 79

Irrigated areas, water availability, water requirements and irrigation potential in the Indian Ocean Coast in South Africa

Sub-basin	Actual irrigated (ha)	Actual water use (km^3/yr)	Water available (km^3/yr)	Irr. wat. requirem. (m^3/ha.yr)	Irrigation potential (ha)
H - S	200 000	2.041	1.906	10 000	190 600
T-W	158 000	1.547	5.898	10 000	589 800
X	68 000	0.681	0.974	10 000	97 400
Total IOC	426 000	4.269	8.778		877 800

For the whole of the Indian Ocean Coast in South Africa the irrigation potential is 877 800 ha, which is more than twice the area irrigated at present, estimated at 426 000 ha [*].

The surface water resources produced in the upper Save basin in **Zimbabwe** are estimated at 4.052 km^3/year, which corresponds to the potential yield [216]. After deducting the amount of water already committed, the quantity of water still available is 2.542 km^3/year. Of this quantity, about one-third, or 0.847 km^3, can be considered as potentially available for the development of irrigation. The surface water resources produced in the upper Buzi and Pungoé basins in Zimbabwe are estimated at 1.024 km^3/year, of which 0.922 km^3/year is still available and one-third of this, or 0.307 km^3/year, for the development of irrigation.

At present 124 804 ha have been developed or planned for irrigation in the upper Save basin [216]. Based on land and water and considering an irrigation water requirement of 11 000 m^3/ha per year according to the present study, it would be possible to irrigate another 77 000 ha [*], which would bring the total to about 201 800 ha. In the upper Buzi and Pungoé basins 7 449 ha have already been developed or planned for irrigation [216] and another 1 750 ha could be developed [*], bringing the total to 9 200 ha. For the whole area in the Indian Ocean Coast this leads to an irrigation potential of 211 000 ha. In the upper Save basin, water is the limiting factor, while in the upper Buzi and Pungoé basins the limiting factor is land.

In **Mozambique** the irrigation potential has been estimated at 128 000 ha in the part situated to the north of the Limpopo basin and at 240 000 ha in the part situated to the south of the Limpopo basin, giving a total of 368 000 ha [159].

TABLE 80
Indian Ocean Coast: irrigation potential, water requirements and areas under irrigation

Country	Irrigation potential (ha)	Gross potential irrigation water requirement per ha (m^3/ha per year)	Gross potential irrigation water requirement total (km^3/year)	Area under irrigation (ha)
Swaziland	93 220	9 000	0.839	67 400
South Africa	877 800	10 000	8.778	688 000
Zimbabwe	211 000	11 000	2.321	65 000
Mozambique	368 000	11 000	4.048	41 710
Sum of countries	1 550 020		15.986	862 110
Total for Ind.Oc.Coast	≤ 1 500 000		15.986	

Problems may arise in the area where the rivers are shared by Swaziland, South Africa and Mozambique. The irrigation potential in that area has been estimated at 93 220 ha for Swaziland, 100 000 ha for South Africa and 240 000 ha for Mozambique. The total of 433 220 ha would require 4.479 km^3/year of irrigation water. The total amount of water flowing to the sea has been estimated at about 6.600 km^3/year.

The East Central Coast

The East Central Coast extends from Mozambique in the south to Somalia in the north. It spreads over five countries and covers 3.4% of the continent (Map 21 and Table 81).

TABLE 81
East Central Coast: areas and rainfall by country

Country	Total area of the country (km^2)	Area of the country within the basin (km^2)	As % of total area of basin (%)	As % of total area of country (%)	Average annual rainfall in the basin area (mm) min.	max.	mean
Malawi	118 480	10 120	1.0	8.5	845	2 305	1 160
Mozambique	801 590	368 879	35.9	46.0	780	1 935	1 140
Tanzania	945 090	434 657	42.4	46.0	395	1 780	965
Kenya	580 370	193 463	18.9	33.3	275	1 615	655
Somalia	637 660	19 133	1.9	3.0	290	435	345
For East Central Coast		1 026 252	100.0		275	2 305	960

Rivers and discharges

The area of Malawi located in the East Central Coast region corresponds to the Lake Chilwa and the Lake Chiuta basins. Both lakes are on the border between Malawi and Mozambique. The average annual runoff in the Lake Chilwa basin is estimated at 1.06 km^3, in the Lake Chiuta basin at 0.61 km^3.

In Mozambique the rivers, except for the Ruvuma, which is the border river between Mozambique and Tanzania, originate from the plateau and mountains within the country, and are usually not perennial. Some of them have important waterfalls and steep slopes. The contribution of the Lugenda River to the Ruvuma River is estimated at about 18 km^3/year. Other important rivers flowing to the sea are the Messalo (3.0 km^3/year at mouth), the Lurio (8.0 km^3/year at mouth), the Ligonha (2.6 km^3/year) and the Licungo (8.9 km^3/year at mouth). This gives a total of 22.5 km^3/year from these rivers alone, which means that the water resources are abundant.

In Tanzania many rivers drain to the coast, the most important being, from the south to the north: Ruvuma, Mbenkuru, Matandu, Rufiji, Ruvu, Wami, Sigi, Msangasi and Pangani. The water resources of Tanzania are quite abundant, but not many figures are available on river discharges. The most important rivers are the Ruvuma on the border between Mozambique and Tanzania with an annual flow to the sea of about 28 km^3, of which the contribution of Tanzania is estimated at 10 km^3, and the Rufiji with an annual runoff of nearly 26 km^3 as measured between 1955 and 1978.

In Kenya two main rivers originate in the East Central Coast. The Tana River originates in the mountains in central Kenya and flows through a semi-arid plain to the sea. It has two seasons of high flooding corresponding to the two rainy seasons. The mean annual runoff is 4.95 km^3, but with a high inter-annual variability. The Athi River is a strongly seasonal river with high flows in April-June and November-December and very low flows in the two intervening seasons. The average annual flow is about 1.80 km^3. The river is characterized by important losses; under low flow conditions, losses of 0.14 km^3/year have been measured over the middle and lower reaches. Effluent discharges from Nairobi make a large contribution to the river flow. Most of the water supply to Nairobi comes from the Tana basin and returns to the Athi basin.

The Lag Badana basin in Somalia is part of the East Central Coast. Surface water resources are rather scarce. Some localized runoff occurs during heavy rainfall, but little water reaches the coast.

Irrigation potential and water requirements

Most of the irrigation potential of **Malawi** is located in the Zambezi basin. Some small-scale development might be possible in the Lake Chilwa basin, for a total of about 1 000 ha [*].

Most of the basin of the Lugenda River in **Mozambique** lies in the Niassa province, where the irrigation potential is estimated at 200 000 ha. The irrigation potential in the Cabo Delgado and Nampula province is estimated at 556 000 ha and in the Zambezia province at 300 000 ha. This gives a total of 856 000 ha [159, *].

For **Tanzania**, the irrigation potential has been identified on the basis of large contiguous areas of land on the major rivers and are therefore not exhaustive. Total irrigation potential in the East Central Coast in Tanzania, 959 360 ha, is detailed below [199].

The Ruvuma and other southern basins:

The Ruvuma River forms the border between Tanzania and Mozambique. The irrigation potential in this zone is limited and has always received low priority at national level. The nature of the topography and drainage patterns of much of the zone render irrigation development of a formal nature difficult and expensive because of the need for flood protection works and pumped irrigation water supplies. The total potential in the southern basins is estimated at 15 240 ha.

The Rufiji basin:

This is the largest river basin in Tanzania. The irrigation potential has been classified under three categories:

- first stage development are those schemes which could be undertaken using runoff water flows and requiring minimal drainage or flood protection works;
- second stage development includes storage/flood control dams and flood protection and drainage works which could be implemented at low cost per unit area developed;
- third stage development includes full control of river flows to allow the maximum possible extension of the irrigable area.

First stage development schemes (total: 34 000 ha) and second stage development schemes (total: 89 000 ha) are located in the Upper Great Ruaha basin and the Kilobero basin, in the upstream part of the Rufiji basin. Of the third stage development schemes, 127 000 ha are located in the upper Great Ruaha basin and 287 000 ha in the Kilombero basin. The remaining 84 800 ha are located in the lower Rufiji basin. In the lower part a large area is covered by the Selous Game Reserve. Water problems may occur in the upper Great Ruaha area (Usanga plains), where the total potential is 207 000 ha, but where the annual flow is probably not more than 2.0 km^3.

The Ruvu basin:

Irrigation development in this basin requires both flood control works and storage for dry season irrigation. The potential ranges from 69 000 to 80 000 ha.

The Wami basin:

The alluvial plains are subject to flooding and any extensive development would require flood control as well as storage for dry season irrigation water. Optimistic estimates of irrigation potential in the alluvial plains range from 40 000 to 48 000 ha, but other estimates are 14 000 ha. In the coastal plains the identified irrigation potential ranges from 37 000 to 44 000 ha.

The Sigi and Msangasi basins:

The irrigation potential in the Sigi basin, by pumping from the river, is estimated at 400 ha. The irrigation potential the Msangasi basin is 4 800 ha, if adequate storage is provided.

The Pangani basin:

About 150 000 ha are estimated to be under traditional irrigation in the upper basin and water availability is a major constraint on future expansion. The remaining potential has been estimated at 21 120 ha and would require storage dams.

In **Kenya**, in the Tana basin, based on mean monthly flow, 132 700 ha could be irrigated. However, based on 80% dependable monthly flow, this area is reduced to 89 200 ha. The area which could be irrigated by renewable groundwater is estimated at 250 ha. This brings the irrigation potential to 89 450 ha. In the Athi basin, based on mean monthly flow, 22 400 ha could be irrigated; based on 80% dependable monthly flow, 21 000 ha. The area which could be irrigated by renewable groundwater is estimated at 650 ha. This brings the irrigation potential to 21 650 ha. The total irrigation potential for the East Central Coast is estimated at 111 100 ha [125].

In view of the scarce water resources in **Somalia**, the irrigation potential has been considered negligible [*].

Table 82 summarizes the figures on irrigation and water requirements for the East Central Coast.

TABLE 82
East Central Coast: irrigation potential, water requirements and areas under irrigation

Country	Irrigation potential (ha)	Gross potential irrigation water requirement		Area under irrigation (ha)
		per ha (m^3/ha per year)	total (km^3/year)	
Malawi	1 000	10 000	0.010	0
Mozambique	856 000	11 000	9.416	5 000
Tanzania	959 360	10 500	10.073	140 000
Kenya	111 100	13 000	1.444	33 610
Somalia	0	8 000	0.000	0
Sum of countries	1 927 460		20.944	178 610
Total for E. Cent. Coast	1 927 460		20.944	

In general, water resources are sufficient for the development of the irrigation potential in the East Central Coast, but problems may arise in the north of Tanzania in the Pangani basin, where water availability is less than required.

The North East Coast

The North East Coast covers 2.4% of the continent and spreads over six countries (Map 22 and Table 83).

TABLE 83
North East Coast: areas and rainfall by country

Country	Total area of the country (km^2)	Area of the country within the basin (km^2)	As % of total area of basin (%)	As % of total area of country (%)	Average annual rainfall in the basin area (mm)		
					min.	max.	mean
Somalia	637 660	392 065	54.0	61.5	0	650	180
Ethiopia	1 100 010	50 173	6.9	4.6	95	725	235
Djibouti	23 200	10 400	1.4	44.8	40	465	145
Eritrea	121 890	88 364	12.2	72.5	40	570	275
Sudan	2 505 810	96 450	13.3	3.8	16	310	80
Egypt	1 001 450	88 250	12.2	8.8	0	135	20
For North East Coast		725 702	100.0		0	725	165

River system and discharges

Five basins can be distinguished in the North East Coast in Somalia:

- In the Gulf of Aden basin the annual upstream runoff is estimated at 0.48 km^3. The quantity of water that disappears by infiltration in the upstream parts is estimated at 0.35 km^3/year, the infiltration at the coastal area at 0.13 km^3/year.
- In the Darror basin there are no significant surface water resources.

- In the Tug Der basin the average annual runoff is estimated at 0.03 km^3. Water flows only after heavy rainfall, but it disappears quickly. Little water reaches the coast.
- In the Ogaden basin surface water resources are scarce due to lack of rainfall.
- The Indian Ocean basin is only a very narrow strip of land along the ocean. The surface drainage is insignificant.

The surface water resources in the Ogaden and Gulf of Aden basins in Ethiopia are considered to be negligible. About 55% of Djibouti drains to the sea to the east. Surface water resources are directly dependent on rainfall (> 10 mm), resulting in rapid floods lasting only a few hours. The internal renewable water resources for the whole of Djibouti are estimated at 0.3 km^3/year.

The Baraka and Anseba rivers rise on the north-western slopes of the central highlands in Eritrea and flow northwards to a confluence near the border with Sudan. Only high rainfall results in flows reaching the Sudanese border, with an average estimated at about 0.8 km^3/year. The Red Sea drainage basin in Eritrea comprises numerous small rivers originating in the eastern escarpment. A global estimate of annual runoff of 0.88 km^3 has been made for the littoral as a whole. The renewable water resources in Egypt are negligible.

Irrigation potential and water requirements

Irrigation potential in **Somalia** can be estimated at about 10 000 ha by spate water at different locations, if dams are constructed [*]. There is no irrigation potential in **Ethiopia** [*].

The cultivable area in **Djibouti** is estimated at about 6 000 ha, but the area equipped for irrigation is only 674 ha, of which about 374 ha are in the North East Coast [93]. No detailed information is available on irrigation potential, but with the available water resources it has been estimated at 1 000 ha, of which 550 ha have been estimated to be in the North East Coast [*].

The land suitable for irrigation in the Barka-Anseba basin in **Eritrea** is about 130 000 ha [100]. It is estimated that, with dam construction, about 6 500 ha can be developed under irrigation. The land suitable for irrigation in the Red Sea drainage basin is about 240 000 ha. It is expected that about 31 000 ha of land lying on the riversbanks can be irrigated.

In **Sudan** about 30 000 ha are expected to be irrigated by spate water [193]. There is no irrigation potential using renewable water resources in **Egypt** [*].

Table 84 summarizes the figures on irrigation and water requirements in the North East Coast.

TABLE 84
North East Coast: irrigation potential, water requirements and areas under irrigation

Country	Irrigation potential (ha)	Gross potential irrigation water requirement		Area under irrigation (ha)
		per ha (m^3/ha per year)	total (km^3/year)	
Somalia	10 000	20 000	0.200	1 000
Ethiopia	0	15 500	0.000	0
Djibouti	550	12 000	0.007	574
Eritrea	37 500	10 500	0.394	13 000
Sudan	30 000	17 000	0.510	10 000
Egypt	0	17 500	0.000	0
Sum of countries	78 050		1.110	24 574
Total for N. East Coast	≤ 78 050		1.110	

The above irrigation potential depends mostly on spate water, which is rather irregular in space and time.

Madagascar

Two major basin groups can be distinguished in Madagascar: the one draining to the west to the Madagascar Channel and the one draining to the east to the Indian Ocean. Rainfall in Madagascar varies from that of tropical rain forest to near desert conditions. The types of irrigation vary according to the three main ecological regions of the country: the Highlands, the West and the narrow East Coast. Because of the high altitude, in the Highlands the dry season (June-October) is cool, which limits crop production. The West is hot and the dry season is very long, up to nine months in the far south-west. Rainfall can be less than 400 mm/year. The East Coast is warm and humid with rainfall that can exceed 3 000 mm/year and with almost no dry season. Irrigation potential has been estimated at 1.5 million hectares and over 70% of this area already benefits from irrigation, although large areas need rehabilitation.

TABLE 85

Madagascar: irrigation potential, water requirements and areas under irrigation

Country	Irrigation potential (ha)	Gross potential irrigation water requirement		Area under irrigation (ha)
		per ha (m^3/ha per year)	total (km^3/year)	
West	1 000 000	16 000	16.000	700 000
East	500 000	14 500	7.250	387 000
For Madagascar	1 500 000		23.250	1 087 000

The renewable water resources are estimated at 337 km^3/year, which is almost 15 times the total water required for the development of the irrigation potential.

Islands

Five countries are grouped in this category, as shown in Table 86.

TABLE 86

Islands: areas and rainfall by country

Country	Total area of the country (km^2)	Average annual rainfall in the basin area (mm)		
		min.	max.	mean
Cape Verde	4 030	60	500	230
Comoros	1 861			900
Mauritius	2 040	700	4 000	2 180
Sao Tome & Principe	960	900	7 000	3 200
Seychelles	455	1 290	2 370	1 740
Total for islands	9 346			

Cape Verde, an island group in the Atlantic Ocean to the west of northern Africa, is a very dry country. The islands of São Tome and Principe are situated in the Gulf of Guinea with very high rainfall. The three other countries are situated in the Indian Ocean to the east of southern Africa. Rainfall varies from an average of 900 mm/year in Comoros to almost 2 200 mm/year in Mauritius. Table 87 summarizes the figures on irrigation and water requirements.

TABLE 87

Islands: irrigation potential, water requirements and areas under irrigation

Country	Irrigation potential (ha)	Gross potential irrigation water requirements		Area under irrigation (ha)
		per ha (m^3/ha per year)	total (km^3/year)	
Cape Verde	2 990	25 000	0.075	2 779
Comoros	300	5 000	0.002	130
Mauritius	20 000	5 000	0.100	17 500
Sao Tome & Principe	10 700	12 500	0.134	9 700
Seychelles	1 000	5 000	0.005	
Total for islands	34 990		0.315	30 109

DATA QUALITY ASSESSMENT

Over 1 000 references have been consulted, of which about 25% contained information that proved to be more or less useful for the purpose of this study, although there is a large variation in the reliability of the information. Table 88 below shows the availability of information on water resources and irrigation potential by country. The first category refers to countries for which reasonably detailed information was available on both water resources and irrigation potential. The second category refers to countries for which either less detailed information was available for both water resources and irrigation potential or reasonable detailed information for one of the two subjects and less for the other. The third category refers to countries for which little information was available on both water resources and irrigation potential.

TABLE 88
Availability of information on water resources and/or irrigation potential by country

Category	Information on water resources and irrigation potential	Countries	Number of countries	% of total number of countries	% of irrigation potential
1	Reasonably detailed	Botswana, Burkina Faso, Cape Verde, Egypt, Kenya, Malawi, Mali, Morocco, Niger, Nigeria, Senegal, South Africa, Swaziland, Tunisia, Zambia, Zimbabwe	16	30	28
2	Less detailed	Algeria, Benin, Chad, Ethiopia, Guinea Bissau, Libya, Madagascar, Mauritania, Mozambique, Namibia, Somalia, Sudan, Tanzania	13	25	31
3	Little information	Angola, Burundi, Cameroon, Central African Republic, Comoros, Congo, Cote d'Ivoire, Djibouti, Equatorial Guinea, Eritrea, Gabon, Gambia, Ghana, Guinea, Lesotho, Liberia, Mauritius, Rwanda, Sao Tome and Principe, Seychelles, Sierra Leone, Togo, Uganda, Zaire	24	45	41
Total			53	100	100

Reasonably detailed information for both water resources and irrigation potential was available for 16 countries out of the total of 53 countries, or 30%. Reasonably detailed information on water resources only existed for 19 countries, on irrigation potential only for 21 countries.

As expected, in general, the least information was available for the more humid countries (category 3).

The table above is related to information available at country level. For the purpose of this study, it was necessary to have information for each basic unit, which refers to the different basin parts within a country. In total there are 136 basic units (Tables 1 and 2). For only 50 basic units, or 36%, reasonably detailed information was available. For 33 basic units, or 24%, partial information was available, while for the remaining 54 basic units, or 40%, no information was available.

In total, 38 countries are located within more than one basin, which means that there is more than one basic unit in the country (see chapter 2 and Table 2). Table 90 shows the availability of information on irrigation potential at the basic unit level for these 38 countries.

TABLE 89
Availability of information on irrigation potential at basic unit level

Degree of availability of information on irrigation potential at basic unit level	Countries with several basic units	Number of countries	% of total number of countries	% of irrigation potential
Reasonably detailed information available for each basic unit of the country	Botswana, Burkina Faso, Ethiopia, Kenya, Mali, Morocco, Niger, Nigeria, South Africa, Tanzania, Tunisia, Zambia, Zimbabwe.	13	34	30
Reasonably detailed information available for part of the basic units of the country	Algeria, Chad, Egypt, Eritrea, Libya, Madagascar, Malawi, Mauritania, Mozambique, Namibia, Senegal, Somalia, Sudan.	13	34	34
Little or no detailed information available at basic unit level of the country	Angola, Benin, Burundi, Cameroon, Central African Republic, Congo, Côte d'Ivoire, Djibouti, Guinea, Rwanda, Uganda, Zaire.	12	32	36
Total		38	100	100

The above table shows, once again, that in general, the more humid countries have the least available information at basic unit level.

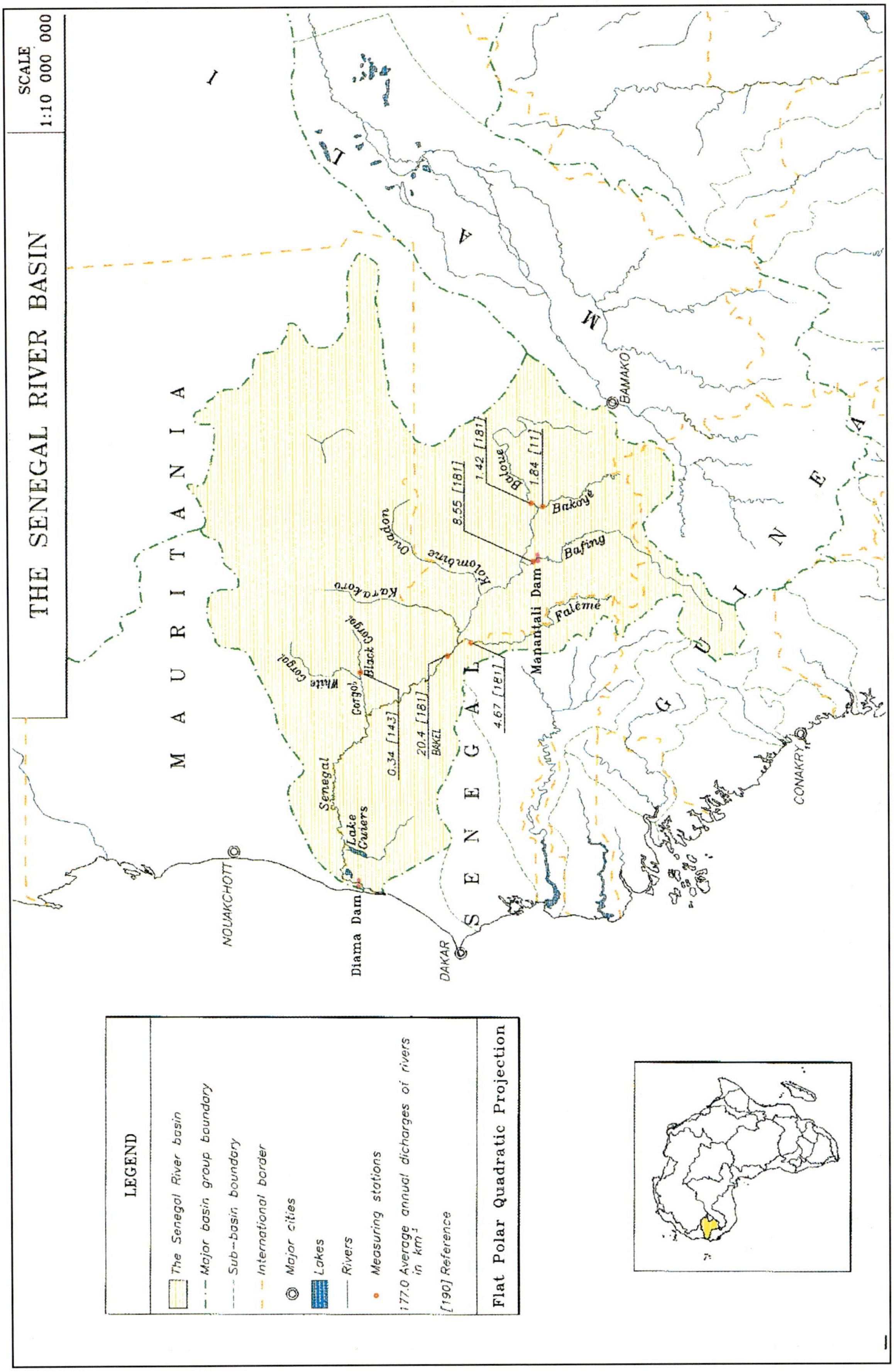
THE SENEGAL RIVER BASIN
SCALE
1:10 000 000
LEGEND
The Senegal River basin
Major basin group boundary
Sub-basin boundary
International border
Major cities
Lakes
Rivers
Measuring stations
177.0 Average annual dicharges of rivers in km³
[190] Reference
Flat Polar Quadratic Projection
MAURITANIA
MALI
SENEGAL
GUINEA
NOUAKCHOTT
DAKAR
BAMAKO
CONAKRY
Diama Dam
Manantali Dam
Lake Guiers
Senegal
Gorgol
Black Gorgol
White Gorgol
Karakoro
Kolombine
Ouadon
Bafing
Bakoye
Baloue
Falémé
0.34 [143]
20.4 [181]
BAKEL
4.67 [181]
8.55 [181]
1.42 [181]
1.84 [11]

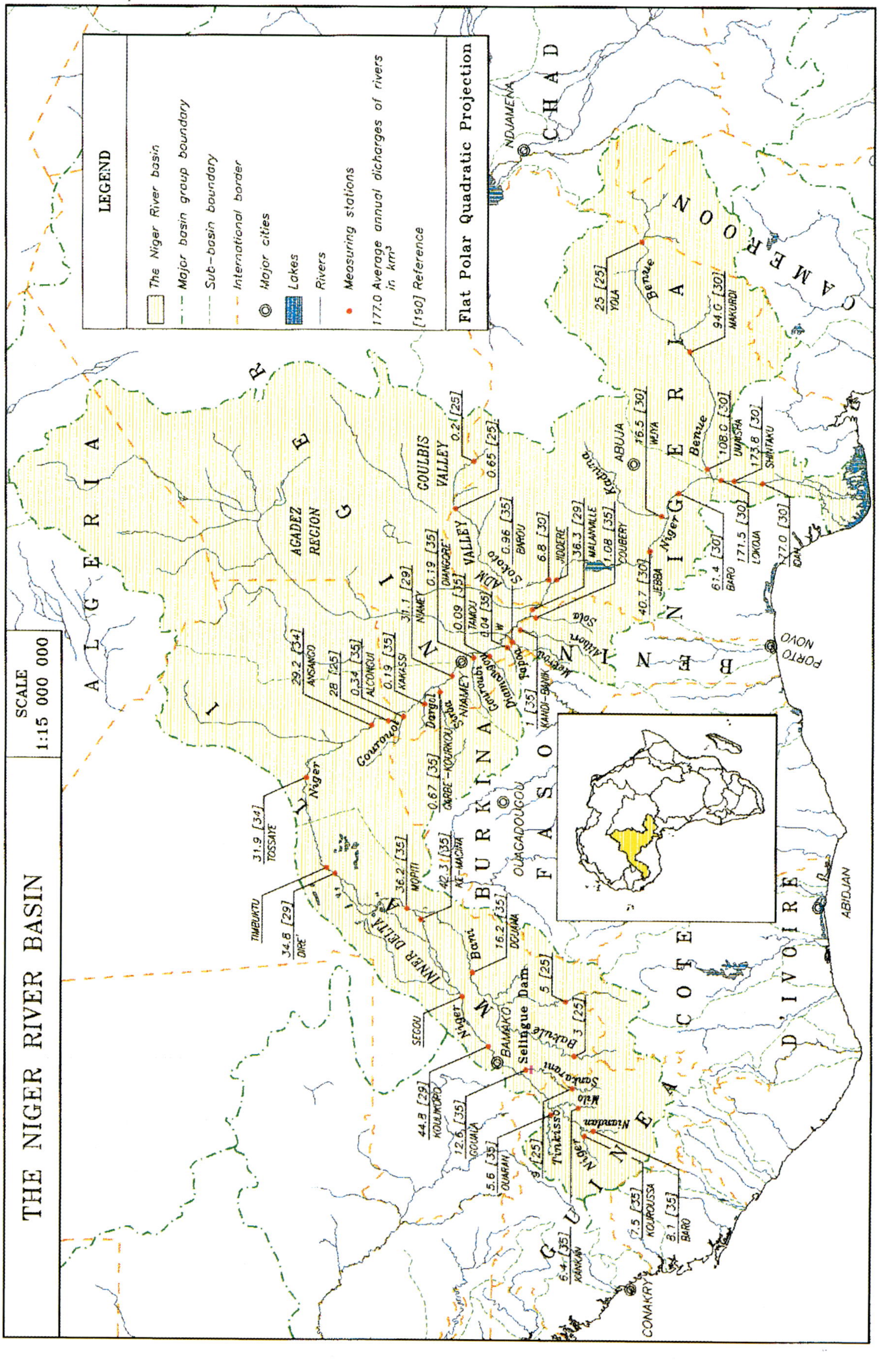
THE NIGER RIVER BASIN
SCALE
1:15 000 000
LEGEND
The Niger River basin
Major basin group boundary
Sub-basin boundary
International border
Major cities
Lakes
Rivers
Measuring stations
177.0 Average annual discharges of rivers in km³
[190] Reference
Flat Polar Quadratic Projection
ALGERIA
NIGER
MALI
BURKINA FASO
NIGERIA
BENIN
CHAD
CAMEROON
GUINEA
COTE D'IVOIRE
AGADEZ REGION
INNER DELTA
NIAMEY
BAMAKO
OUAGADOUGOU
ABUJA
NDJAMENA
CONAKRY
ABIDJAN
PORTO NOVO
Selingue Dam

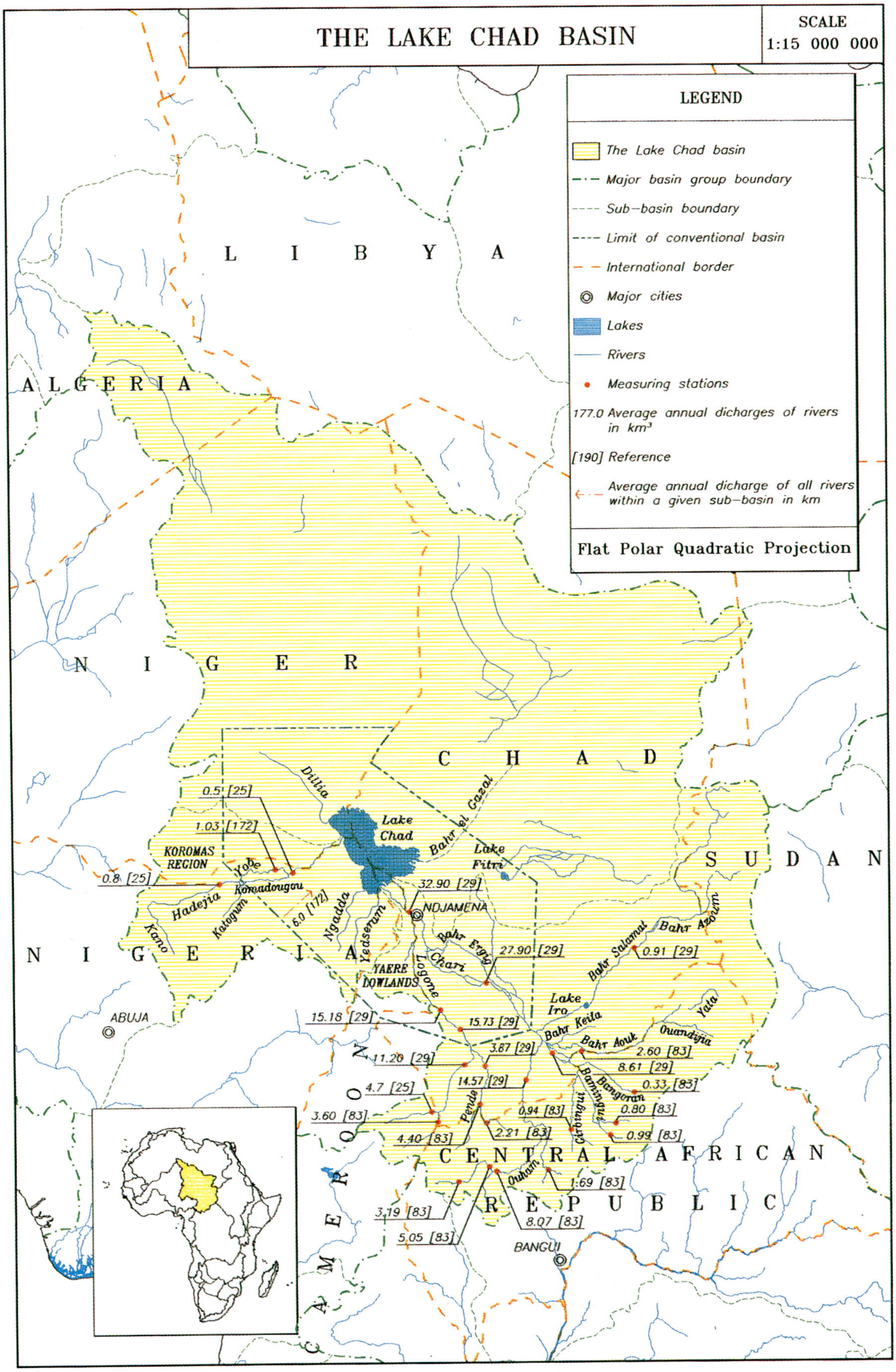
THE LAKE CHAD BASIN
SCALE
1:15 000 000
LEGEND
The Lake Chad basin
Major basin group boundary
Sub-basin boundary
Limit of conventional basin
International border
Major cities
Lakes
Rivers
Measuring stations
177.0 Average annual dicharges of rivers in km³
[190] Reference
Average annual dicharge of all rivers within a given sub-basin in km
Flat Polar Quadratic Projection
LIBYA
ALGERIA
NIGER
CHAD
SUDAN
NIGERIA
CAMEROON
CENTRAL AFRICAN REPUBLIC
Lake Chad
Lake Fitri
Lake Iro
NDJAMENA
ABUJA
BANGUI
KOROMAS REGION
YAERE LOWLANDS
Dillia
Bahr el Gazal
Yobe
Komadougou
Hadejia
Katagum
Kano
Ngadda
Yedseram
Logone
Chari
Bahr Erguig
Bahr Salamat
Bahr Azoum
Bahr Keita
Bahr Aouk
Ouandjia
Yata
Bangoran
Bamingui
Gribingui
Ouham
Pende
0.5 [25]
1.03 [172]
0.8 [25]
6.0 [172]
32.90 [29]
27.90 [29]
0.91 [29]
15.18 [29]
15.73 [29]
3.87 [29]
11.20 [29]
14.57 [29]
4.7 [25]
3.60 [83]
4.40 [83]
0.94 [83]
2.21 [83]
2.60 [83]
8.61 [29]
0.33 [83]
0.80 [83]
0.99 [83]
1.69 [83]
3.19 [83]
5.05 [83]
8.07 [83]

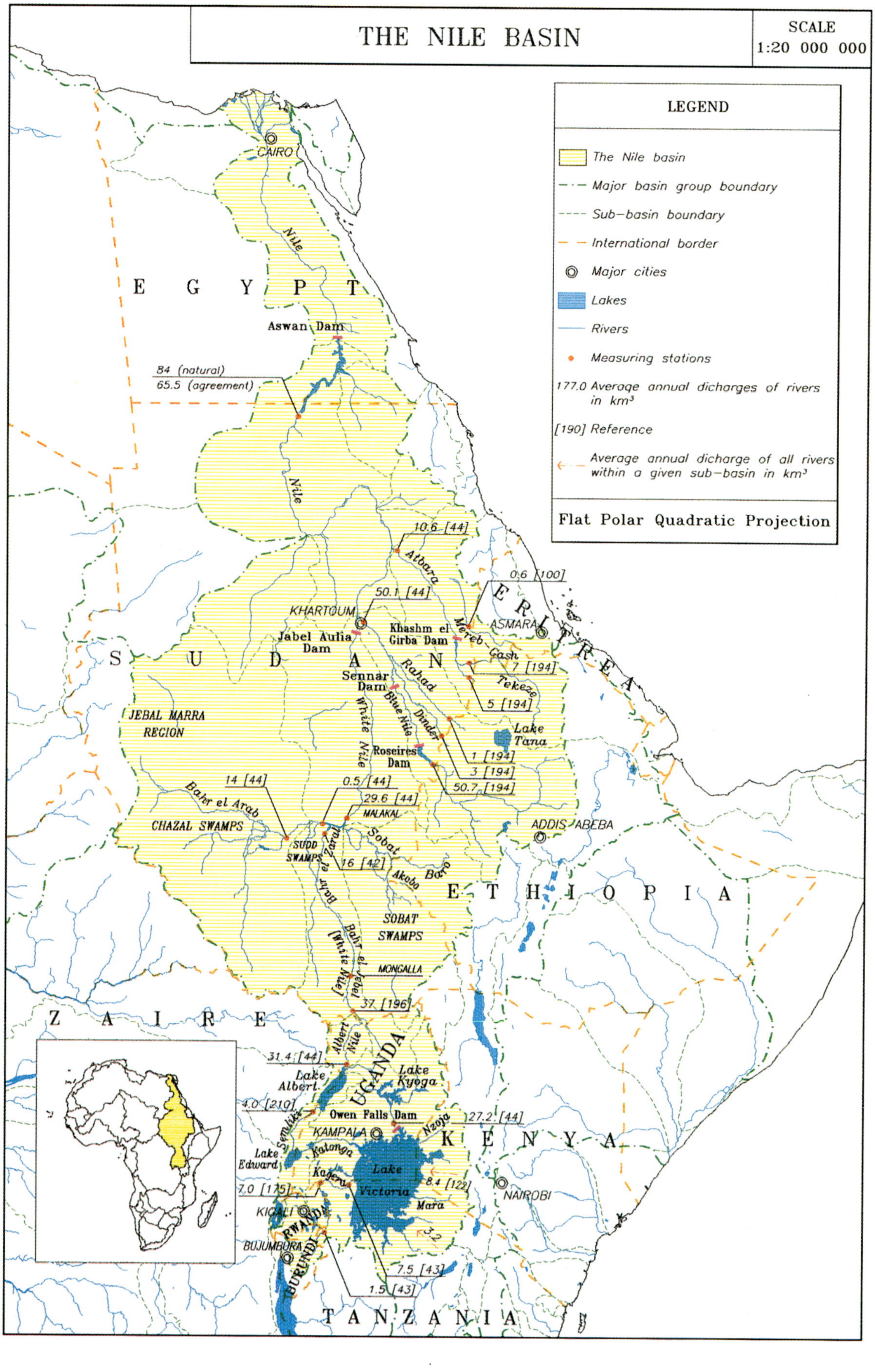
THE NILE BASIN
SCALE
1:20 000 000
LEGEND
The Nile basin
Major basin group boundary
Sub-basin boundary
International border
Major cities
Lakes
Rivers
Measuring stations
177.0 Average annual dicharges of rivers in km³
[190] Reference
Average annual dicharge of all rivers within a given sub-basin in km³
Flat Polar Quadratic Projection
EGYPT
CAIRO
Nile
Aswan Dam
84 (natural)
65.5 (agreement)
SUDAN
KHARTOUM
Jabel Aulia Dam
Khashm el Girba Dam
Sennar Dam
Roseires Dam
Atbara
10.6 [44]
50.1 [44]
ERITREA
ASMARA
0.6 [100]
Mereb
Gash
7 [194]
Tekeze
5 [194]
Rahad
Dinder
Blue Nile
White Nile
Lake Tana
1 [194]
3 [194]
50.7 [194]
JEBAL MARRA REGION
14 [44]
Bahr el Arab
CHAZAL SWAMPS
0.5 [44]
29.6 [44]
MALAKAL
Sobat
Zaraf
SUDD SWAMPS
16 [42]
Akobo
Baro
ADDIS ABEBA
ETHIOPIA
SOBAT SWAMPS
Bahr el Jebel
[White Nile]
MONGALLA
37 [196]
ZAIRE
Albert Nile
UGANDA
31.4 [44]
Lake Albert
Lake Kyoga
4.0 [210]
Semliki
Owen Falls Dam
27.2 [44]
KAMPALA
Nzoia
KENYA
Katonga
Lake Edward
Kagera
Lake Victoria
7.0 [175]
8.4 [122]
NAIROBI
KIGALI
RWANDA
Mara
BUJUMBURA
BURUNDI
7.5 [43]
1.5 [43]
TANZANIA

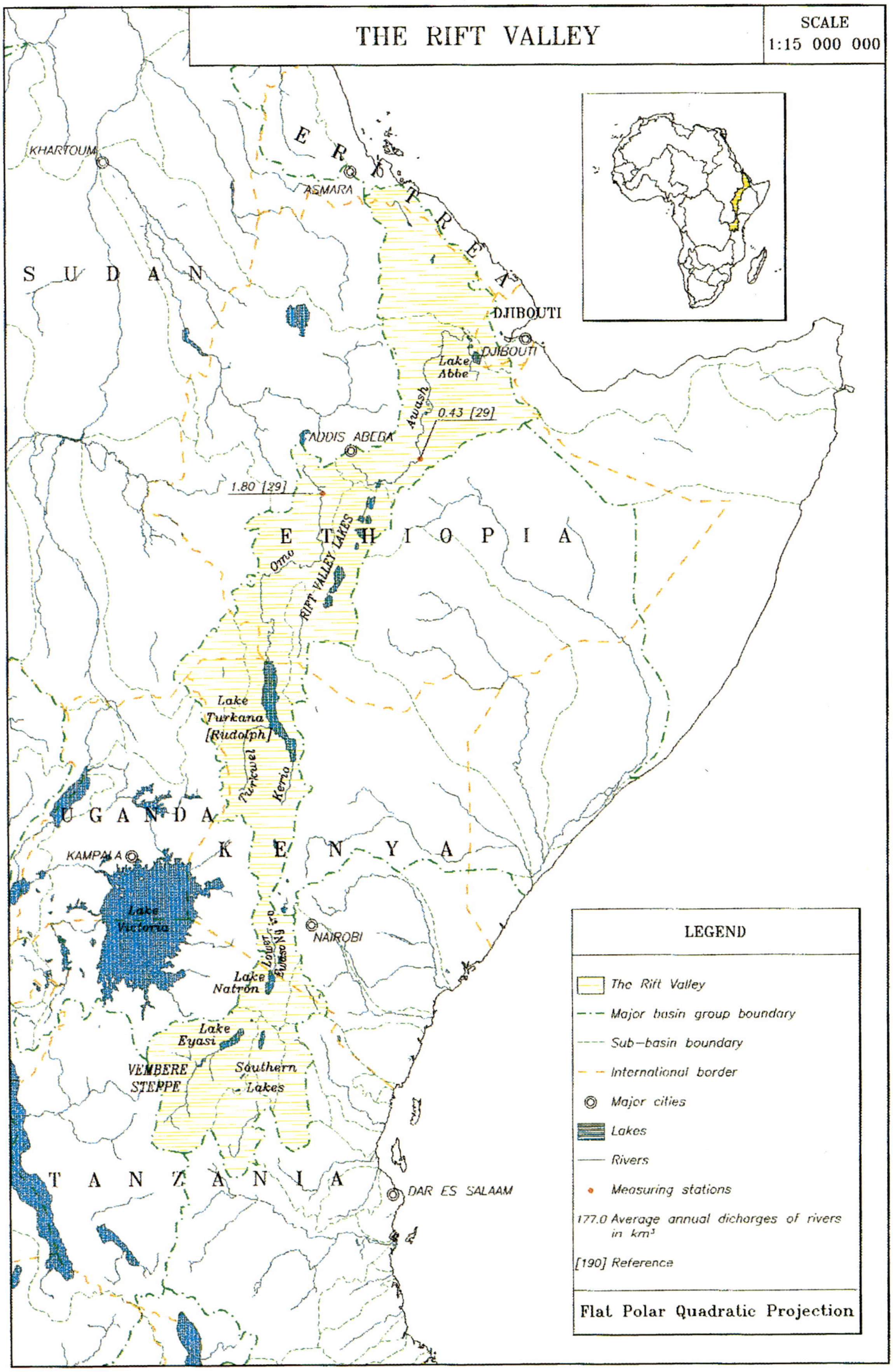
THE RIFT VALLEY
SCALE
1:15 000 000
KHARTOUM
ASMARA
ERITREA
SUDAN
DJIBOUTI
DJIBOUTI
Lake Abbe
Awash
0.43 [29]
ADDIS ABEBA
1.80 [29]
ETHIOPIA
Omo
RIFT VALLEY LAKES
Lake Turkana [Rudolph]
Turkwel
Kerio
UGANDA
KAMPALA
KENYA
Lake Victoria
NAIROBI
Lake Natron
Lake Eyasi
VEMBERE STEPPE
Southern Lakes
TANZANIA
DAR ES SALAAM
LEGEND
The Rift Valley
Major basin group boundary
Sub-basin boundary
International border
Major cities
Lakes
Rivers
Measuring stations
177.0 Average annual dicharges of rivers in km³
[190] Reference
Flat Polar Quadratic Projection

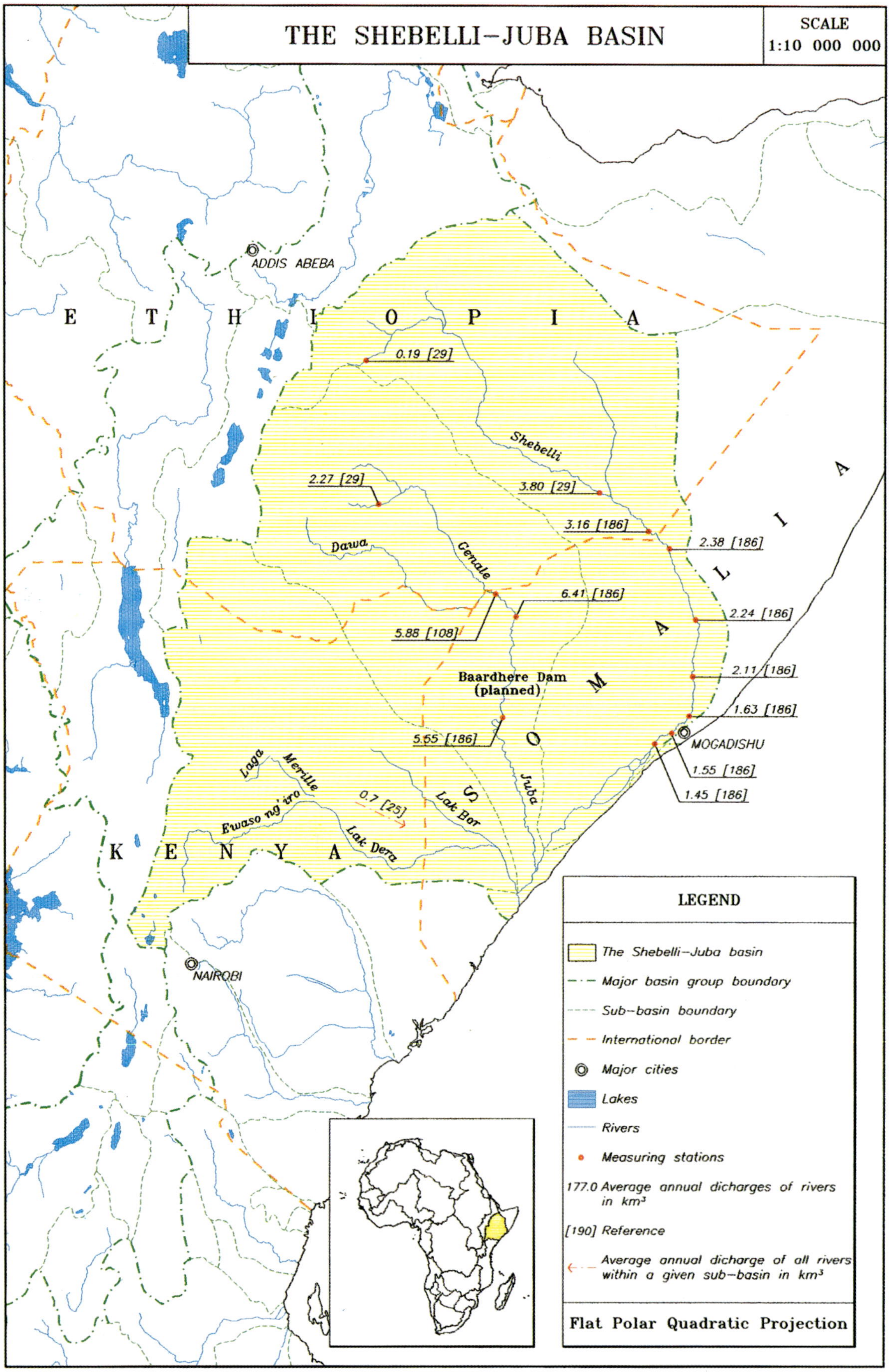
THE SHEBELLI–JUBA BASIN
SCALE
1:10 000 000
ADDIS ABEBA
E T H I O P I A
S O M A L I A
K E N Y A
NAIROBI
MOGADISHU
Shebelli
Dawa
Genale
Juba
Laga
Merille
Ewaso ng'iro
Lak Bor
Lak Dera
Baardhere Dam
(planned)
0.19 [29]
2.27 [29]
3.80 [29]
3.16 [186]
2.38 [186]
6.41 [186]
5.88 [108]
2.24 [186]
2.11 [186]
1.63 [186]
5.55 [186]
1.55 [186]
1.45 [186]
0.7 [25]
LEGEND
The Shebelli–Juba basin
Major basin group boundary
Sub–basin boundary
International border
Major cities
Lakes
Rivers
Measuring stations
177.0 Average annual dicharges of rivers in km³
[190] Reference
Average annual dicharge of all rivers within a given sub–basin in km³
Flat Polar Quadratic Projection

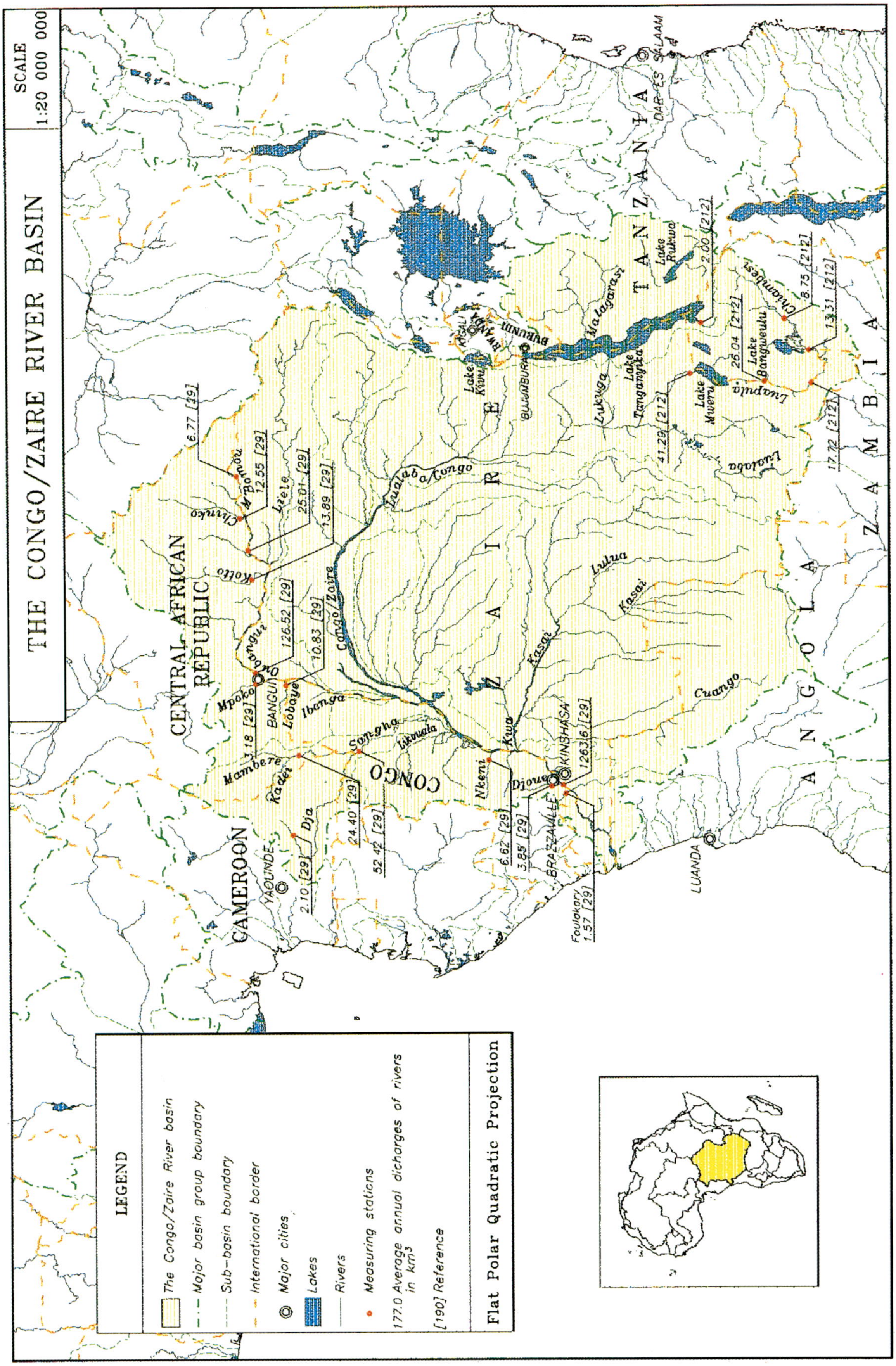
THE CONGO/ZAIRE RIVER BASIN
SCALE
1:20 000 000
LEGEND
The Congo/Zaire River basin
Major basin group boundary
Sub-basin boundary
International border
Major cities
Lakes
Rivers
Measuring stations
177.0 Average annual dicharges of rivers in km3
[190] Reference
Flat Polar Quadratic Projection
CENTRAL AFRICAN REPUBLIC
CAMEROON
CONGO
ZAIRE
TANZANIA
ZAMBIA
ANGOLA
BURUNDI
RWANDA
BANGUI
YAOUNDE
BRAZZAVILLE
KINSHASA
LUANDA
KIGALI
BUJUMBURA
DAR-ES-SALAAM
Lake Kivu
Lake Tanganyika
Lake Mweru
Lake Bangweulu
Lake Rukwa
Congo/Zaire
Lualaba/Congo
Lualaba
Luapula
Chambesi
Lukuga
Malagarasi
Ubangui
Mpoko
Lobaye
Ibanga
Sangha
Likouala
Mambere
Kadei
Dja
Nkeni
Kwa
Djoue
Kasai
Lulua
Cuango
Kotto
Chinko
M'Bomou
Uele
6.77 [29]
12.55 [29]
25.01 [29]
13.89 [29]
126.52 [29]
10.83 [29]
3.18 [29]
24.40 [29]
52.42 [29]
2.10 [29]
6.62 [29]
3.85 [29]
1263.6 [29]
Foulakary 1.57 [29]
41.23 [212]
26.04 [212]
8.75 [212]
1.31 [212]
2.00 [212]
17.72 [212]

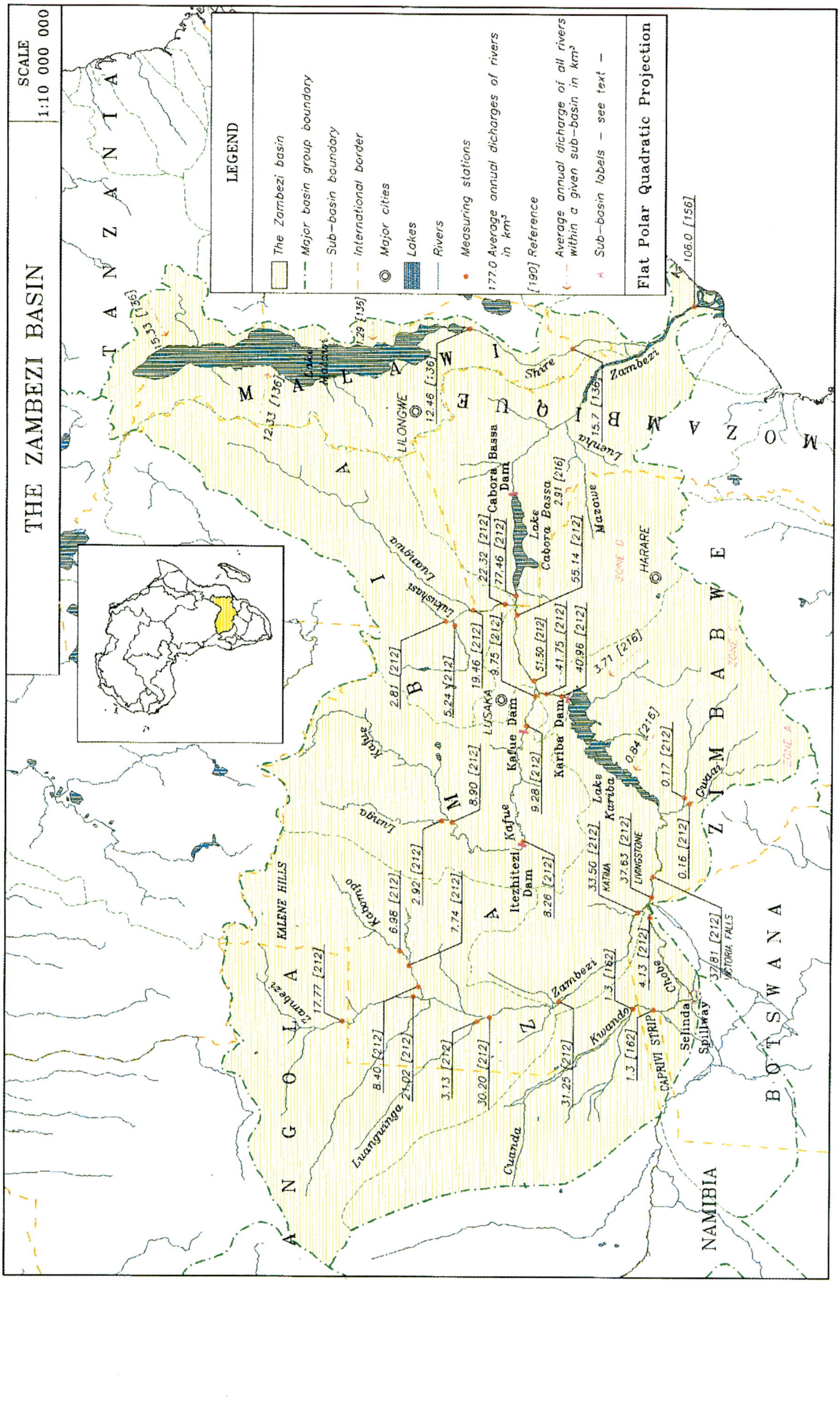
THE ZAMBEZI BASIN
SCALE
1:10 000 000
LEGEND
The Zambezi basin
Major basin group boundary
Sub-basin boundary
International border
Major cities
Lakes
Rivers
Measuring stations
177.0 Average annual dicharges of rivers in km³
[190] Reference
Average annual dicharge of all rivers within a given sub-basin in km³
Sub-basin labels - see text -
Flat Polar Quadratic Projection
TANZANIA
MALAWI
ZAMBIA
MOZAMBIQUE
ZIMBABWE
BOTSWANA
NAMIBIA
ANGOLA
Lake Malawi
LILONGWE
LUSAKA
HARARE
Cabora Bassa Dam
Lake Cabora Bassa
Kafue Dam
Kariba Dam
Lake Kariba
Itezhitezi Dam
KALENE HILLS
CAPRIVI STRIP
Selinda Spillway
Zambezi
Shire
Luenha
Mazowe
Luangwa
Lunsemfwa
Kafue
Lunga
Kabompo
Luanginga
Cuando
Kwando
Chobe
Gwaai
VICTORIA FALLS
LIVINGSTONE
KATIMA
106.0 [156]
15.7 [136]
12.46 [136]
12.33 [136]
15.33 [136]
1.29 [136]
22.32 [212]
77.46 [212]
2.91 [216]
55.14 [212]
9.75 [212]
19.46 [212]
5.24 [212]
2.81 [212]
51.50 [212]
41.75 [212]
40.96 [212]
3.71 [216]
8.90 [212]
9.28 [212]
0.84 [216]
0.17 [212]
0.16 [212]
8.26 [212]
6.98 [212]
2.92 [212]
7.74 [212]
33.50 [212]
37.63 [212]
37.81 [212]
17.77 [212]
8.40 [212]
21.02 [212]
3.13 [212]
30.20 [212]
31.25 [212]
1.3 [162]
4.13 [212]
ZONE A
ZONE C
ZONE D

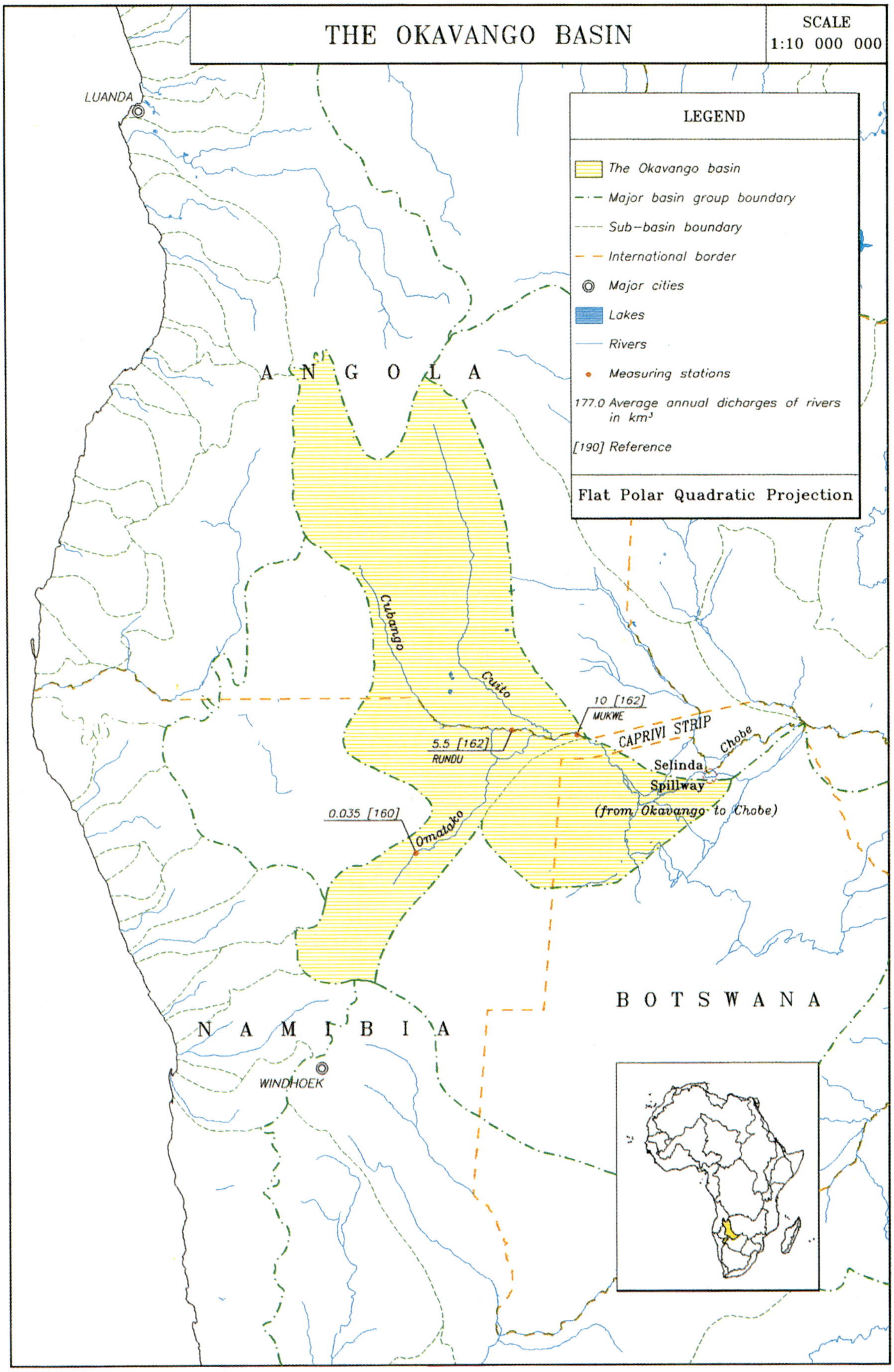
THE OKAVANGO BASIN
SCALE
1:10 000 000
LEGEND
The Okavango basin
Major basin group boundary
Sub-basin boundary
International border
Major cities
Lakes
Rivers
Measuring stations
177.0 Average annual dicharges of rivers in km³
[190] Reference
Flat Polar Quadratic Projection
LUANDA
A N G O L A
Cubango
Cuito
10 [162]
MUKWE
5.5 [162]
RUNDU
CAPRIVI STRIP
Chobe
Selinda
Spillway
(from Okavango to Chobe)
0.035 [160]
Omatako
BOTSWANA
N A M I B I A
WINDHOEK

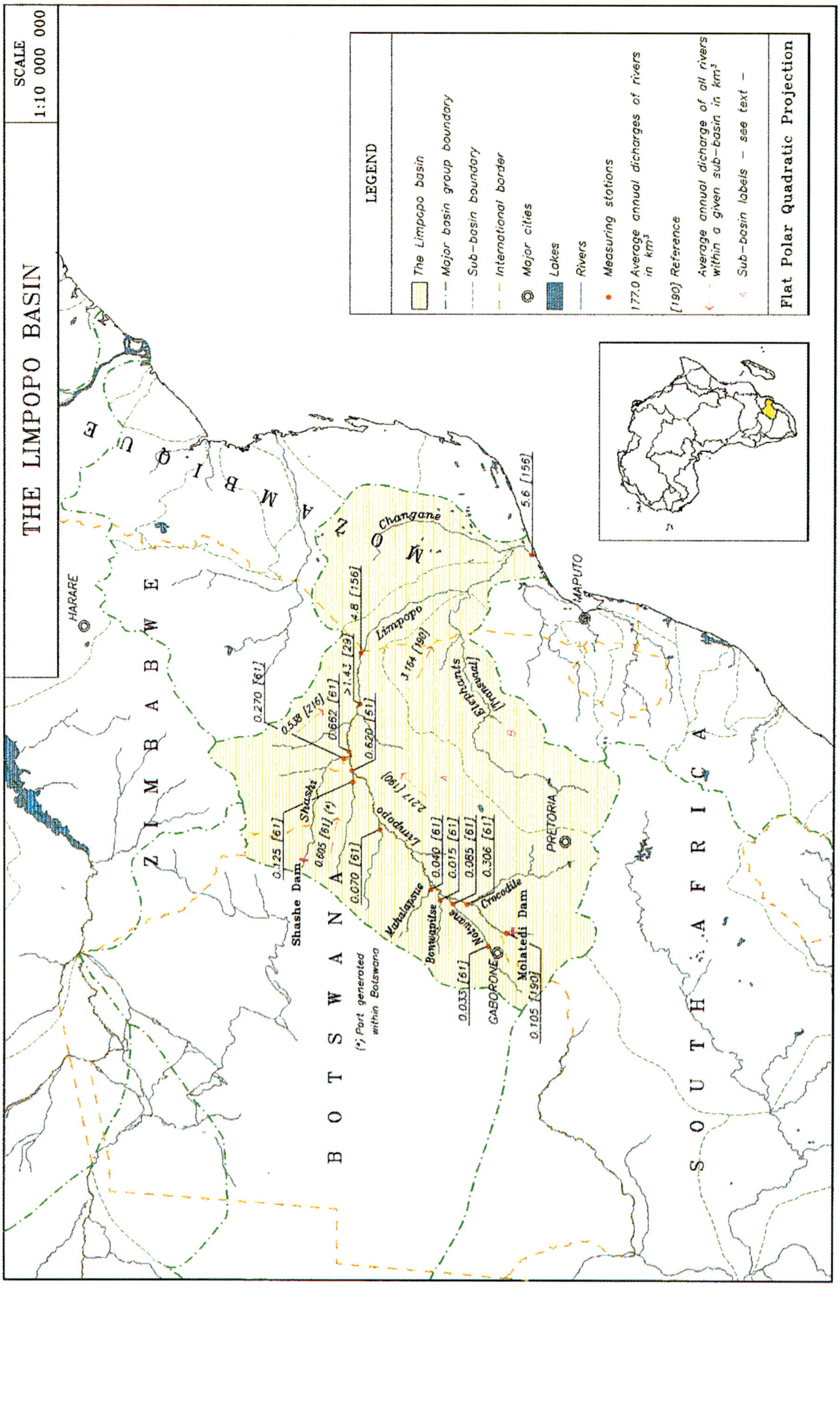
THE LIMPOPO BASIN
SCALE
1:10 000 000
LEGEND
The Limpopo basin
Major basin group boundary
Sub-basin boundary
International border
Major cities
Lakes
Rivers
Measuring stations
177.0 Average annual dicharges of rivers in km³
[190] Reference
Average annual dicharge of all rivers within a given sub-basin in km³
Sub-basin labels - see text -
Flat Polar Quadratic Projection
ZIMBABWE
MOZAMBIQUE
BOTSWANA
SOUTH AFRICA
HARARE
MAPUTO
PRETORIA
GABORONE
Shashe Dam
Molatedi Dam
Limpopo
Changane
Shashi
Elephants
(Olifants)
Crocodile
Notwane
Bonwapitse
Mahalapswe
(*) Part generated within Botswana
5.6 [156]
4.8 [156]
>1.43 [29]
0.620 [51]
0.652 [61]
0.538 [216]
0.270 [61]
0.125 [61]
0.605 [61] (*)
0.070 [61]
0.040 [61]
0.015 [61]
0.085 [61]
0.306 [61]
0.033 [61]
0.105 [190]
2.217 [190]
3.164 [190]

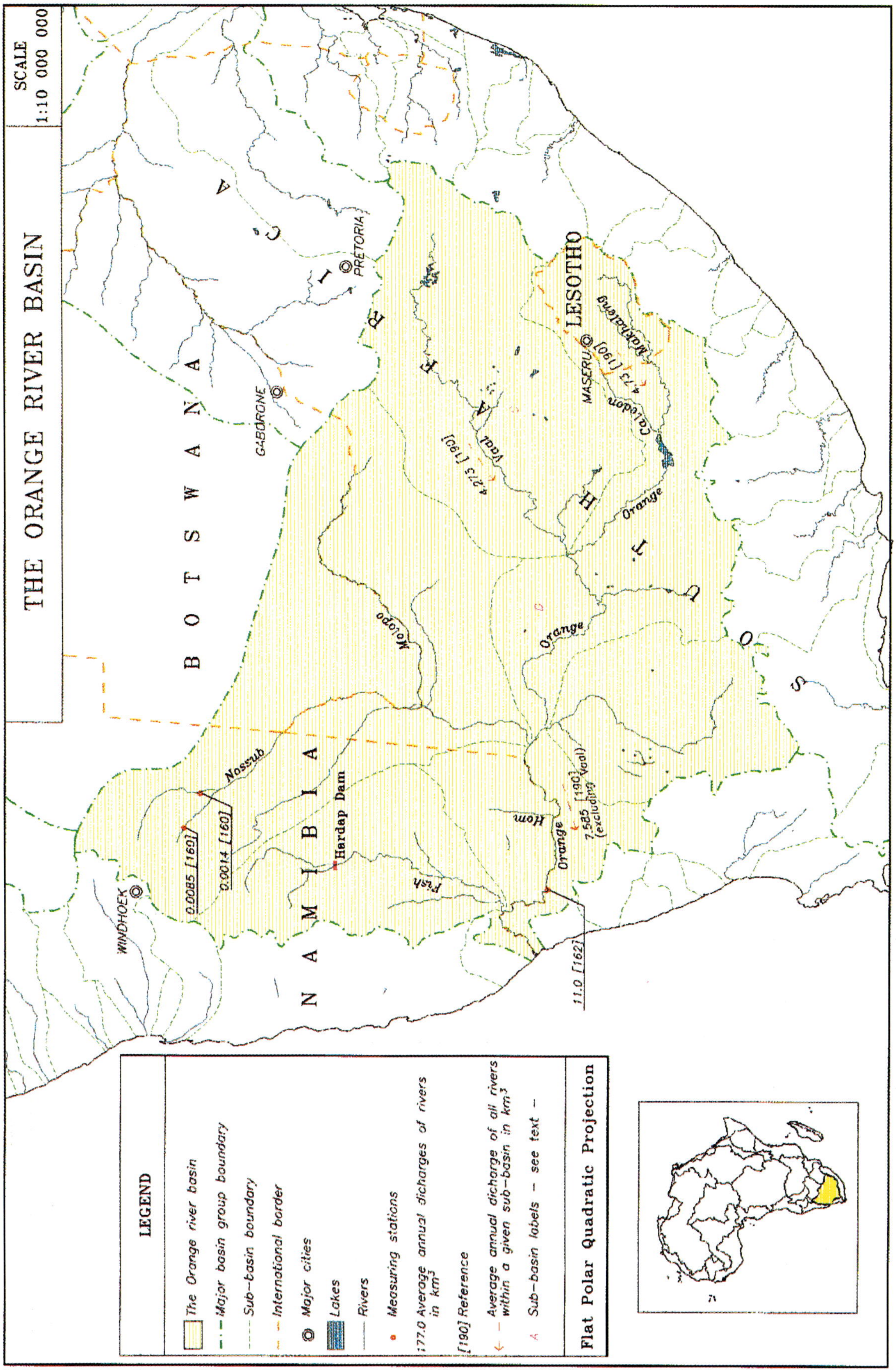
THE ORANGE RIVER BASIN
SCALE
1:10 000 000
BOTSWANA
NAMIBIA
LESOTHO
AFRICA
SOUTH
PRETORIA
GABORONE
WINDHOEK
MASERU
Hardap Dam
Orange
Vaal
Caledon
Makhaleng
Molopo
Nossub
Fish
Hom
4.273 [190]
4.73 [190]
0.0085 [160]
0.0014 [160]
7.585 [190] (excluding Vaal)
11.0 [162]
LEGEND
The Orange river basin
Major basin group boundary
Sub-basin boundary
International border
Major cities
Lakes
Rivers
Measuring stations
177.0 Average annual discharges of rivers in km3
[190] Reference
Average annual discharge of all rivers within a given sub-basin in km3
Sub-basin labels - see text -
Flat Polar Quadratic Projection

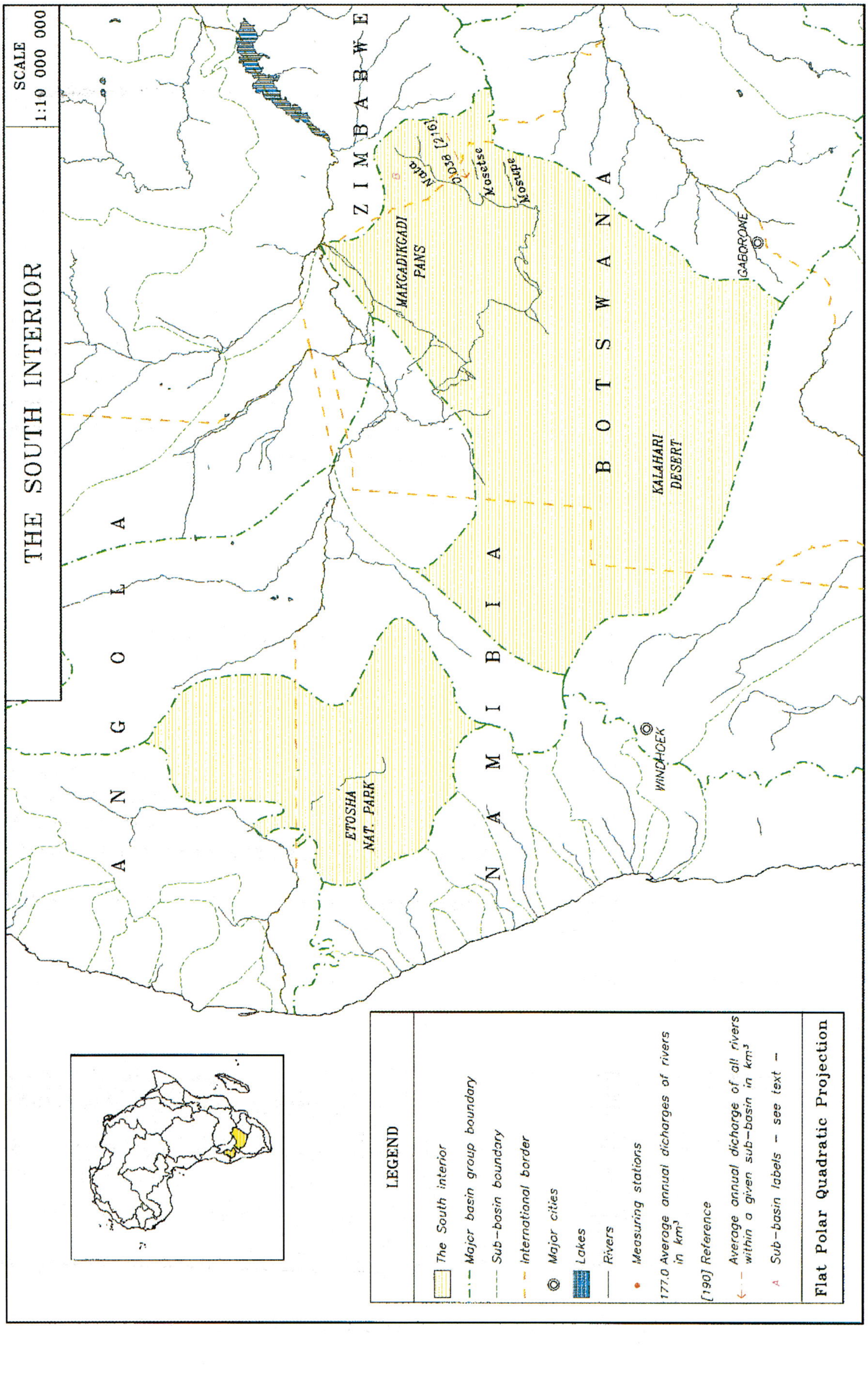
THE SOUTH INTERIOR
SCALE
1:10 000 000
A N G O L A
Z I M B A B W E
B O T S W A N A
N A M I B I A
MAKGADIKGADI PANS
Nata
0.038 [216]
Mosetse
Mosupe
GABORONE
KALAHARI DESERT
ETOSHA NAT. PARK
WINDHOEK
LEGEND
The South interior
Major basin group boundary
Sub-basin boundary
International border
Major cities
Lakes
Rivers
Measuring stations
177.0 Average annual dicharges of rivers in km³
[190] Reference
Average annual dicharge of all rivers within a given sub-basin in km³
Sub-basin labels - see text -
Flat Polar Quadratic Projection

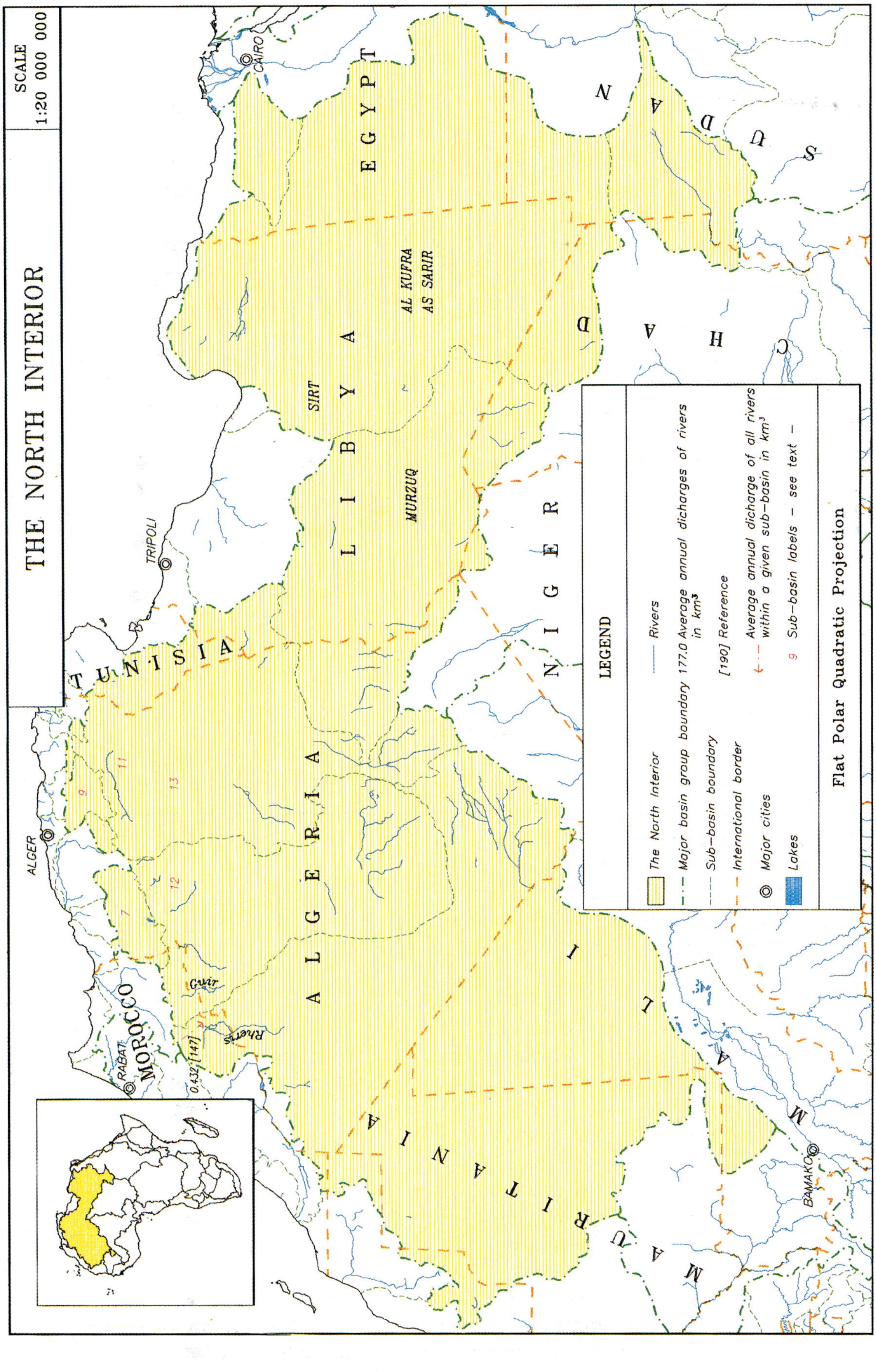
THE NORTH INTERIOR
SCALE
1:20 000 000
LEGEND
The North Interior
Major basin group boundary
Sub-basin boundary
International border
Major cities
Lakes
Rivers
177.0 Average annual dicharges of rivers in km³
[190] Reference
Average annual dicharge of all rivers within a given sub-basin in km³
9 Sub-basin labels – see text –
Flat Polar Quadratic Projection
EGYPT
LIBYA
TUNISIA
ALGERIA
MOROCCO
NIGER
CHAD
SUDAN
MALI
MAURITANIA
AL KUFRA
AS SARIR
SIRT
MURZUQ
CAIRO
TRIPOLI
ALGER
RABAT
BAMAKO
Guir
Rheris
0.432 [147]
9
11
13
12
7

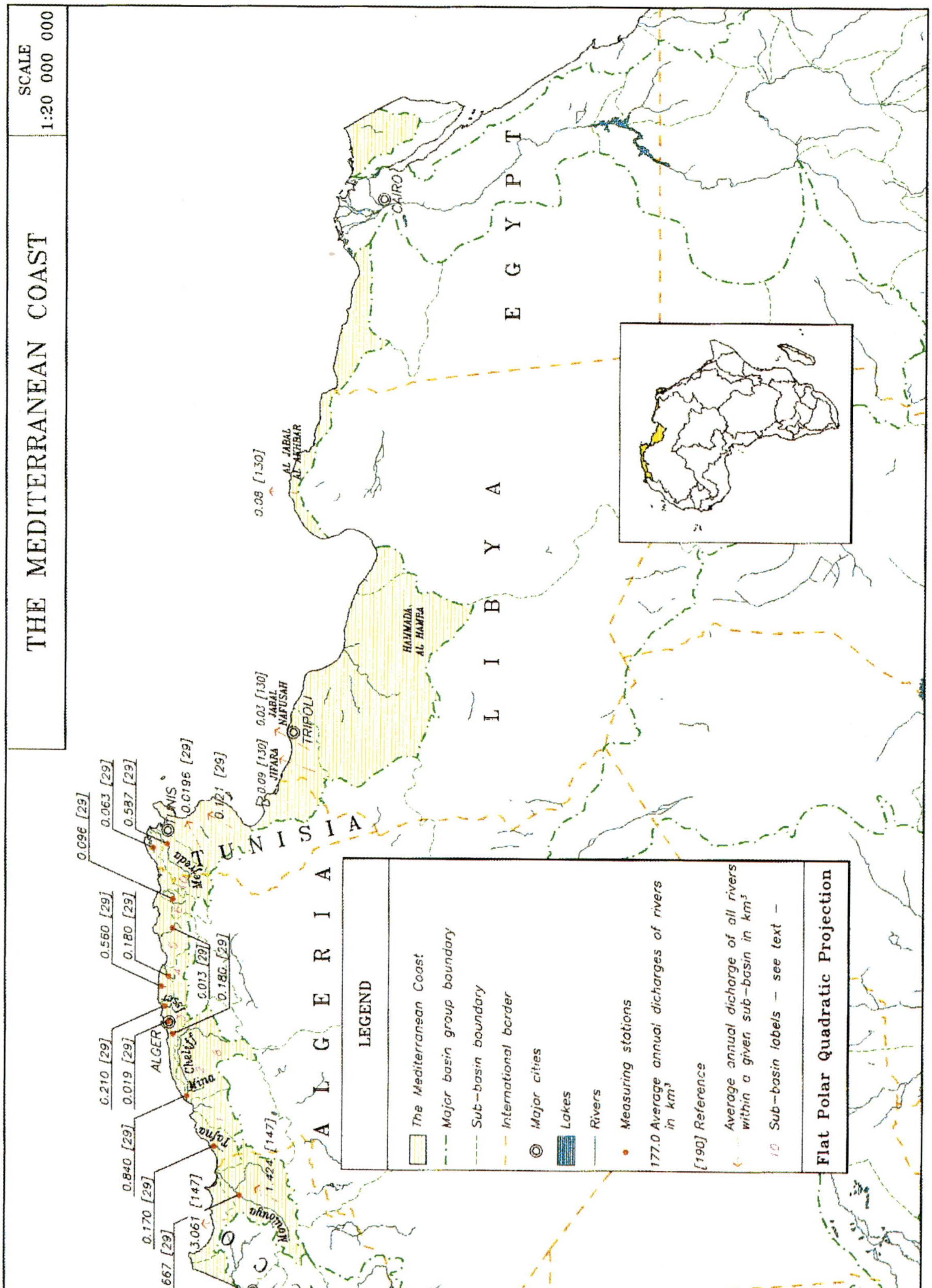
THE MEDITERRANEAN COAST
SCALE
1:20 000 000
LEGEND
The Mediterranean Coast
Major basin group boundary
Sub-basin boundary
International border
Major cities
Lakes
Rivers
Measuring stations
177.0 Average annual discharges of rivers in km³
[190] Reference
Average annual discharge of all rivers within a given sub-basin in km³
Sub-basin labels - see text -
Flat Polar Quadratic Projection
MOROCCO
ALGERIA
TUNISIA
LIBYA
EGYPT
RABAT
ALGER
TUNIS
TRIPOLI
CAIRO
JIFARA
JABAL NAFUSAH
HAMMADA AL HAMRA
AL JABAL AL AKHDAR
Moulouya
Tafna
Mina
Cheliff
Medjerda
0.667 [29]
3.061 [147]
0.170 [29]
0.840 [29]
1.424 [147]
0.210 [29]
0.019 [29]
0.560 [29]
0.180 [29]
0.013 [29]
0.180 [29]
0.096 [29]
0.063 [29]
0.587 [29]
0.0196 [29]
0.121 [29]
0.09 [130]
0.03 [130]
0.08 [130]

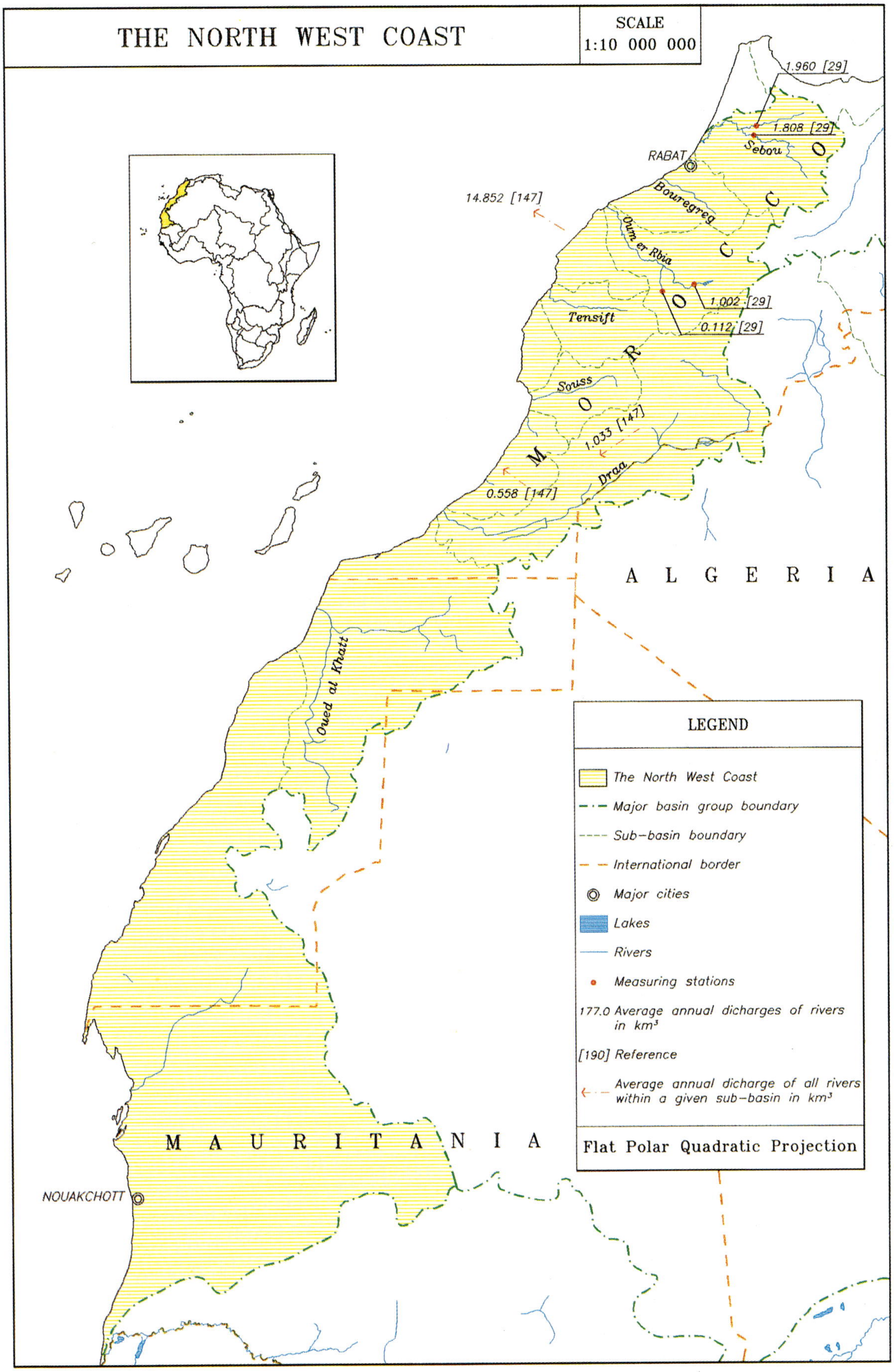
THE NORTH WEST COAST
SCALE
1:10 000 000
1.960 [29]
1.808 [29]
Sebou
RABAT
14.852 [147]
Bouregreg
Oum er Rbia
1.002 [29]
0.112 [29]
Tensift
Souss
1.033 [147]
Draa
0.558 [147]
M O R O C C O
A L G E R I A
Oued al Khatt
M A U R I T A N I A
NOUAKCHOTT
LEGEND
The North West Coast
Major basin group boundary
Sub-basin boundary
International border
Major cities
Lakes
Rivers
Measuring stations
177.0 Average annual dicharges of rivers in km³
[190] Reference
Average annual dicharge of all rivers within a given sub-basin in km³
Flat Polar Quadratic Projection

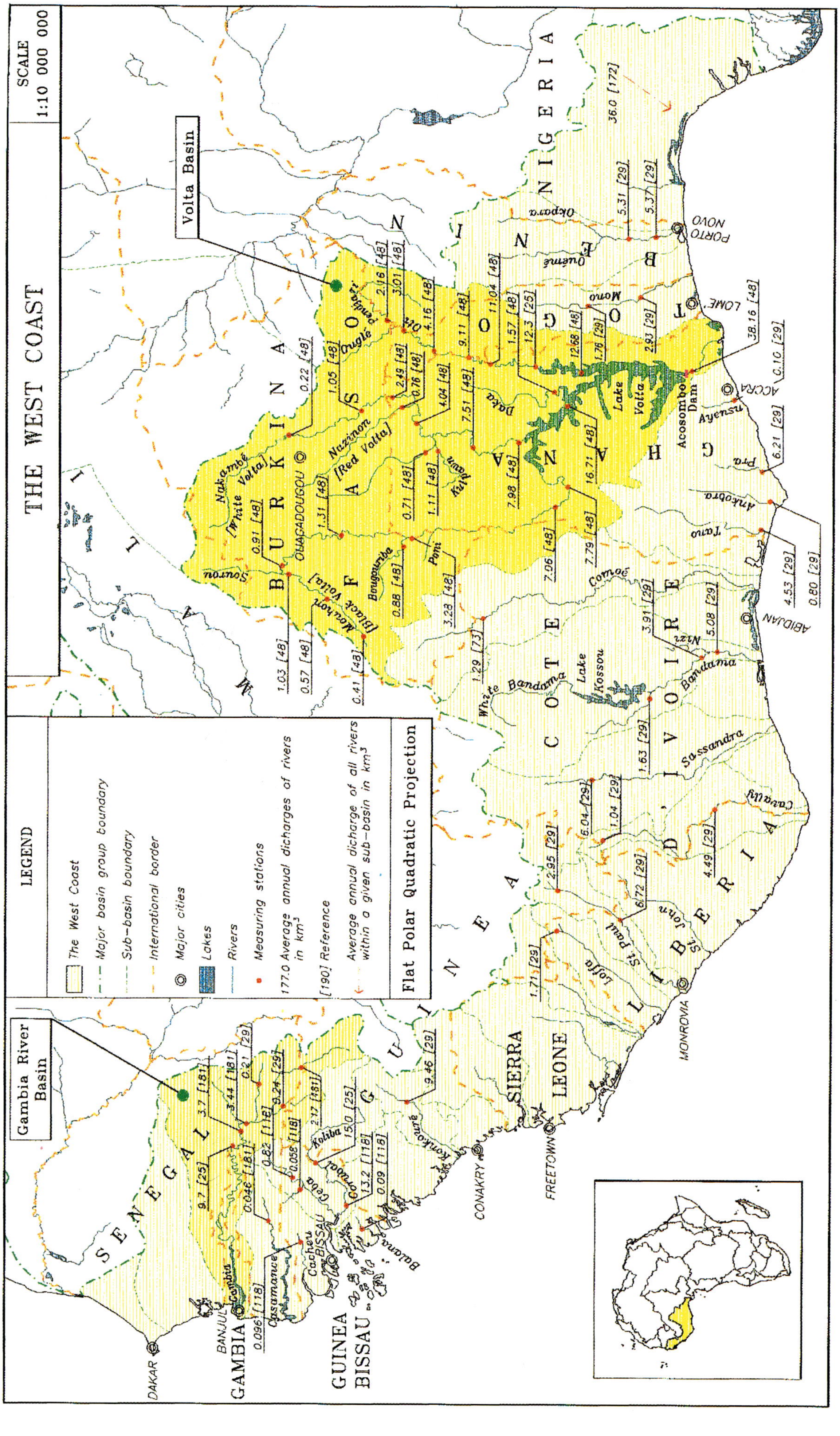
THE WEST COAST
SCALE
1:10 000 000
LEGEND
The West Coast
Major basin group boundary
Sub-basin boundary
International border
Major cities
Lakes
Rivers
Measuring stations
177.0 Average annual dicharges of rivers in km3
[190] Reference
Average annual dicharge of all rivers within a given sub-basin in km3
Flat Polar Quadratic Projection
Gambia River Basin
Volta Basin
SENEGAL
GAMBIA
GUINEA BISSAU
GUINEA
SIERRA LEONE
LIBERIA
COTE D'IVOIRE
GHANA
TOGO
BENIN
NIGERIA
BURKINA FASO
MALI
NIGER
DAKAR
BANJUL
BISSAU
CONAKRY
FREETOWN
MONROVIA
ABIDJAN
ACCRA
LOME
PORTO NOVO
OUAGADOUGOU
Lake Volta
Acosombo Dam
Lake Kossou
Gambia
Casamance
Cacheu
Geba
Corubal
Koliba
Konkouré
Balana
Loffa
St Paul
St. John
Cavally
Sassandra
Bandama
White Bandama
Nzi
Comoé
Tano
Ankobra
Pra
Ayensu
Mono
Oueme
Okpara
Nakambé
[White Volta]
Nazinon
[Red Volta]
Mouhoun
[Black Volta]
Sourou
Bougouriba
Poni
Kulpawn
Daka
Oti
Pendjari
Ougle
36.0 [172]
5.31 [29]
5.37 [29]
38.16 [48]
0.10 [29]
6.21 [29]
4.53 [29]
0.80 [29]
11.04 [48]
1.57 [48]
12.3 [25]
12.68 [48]
1.76 [29]
2.93 [29]
9.11 [48]
2.16 [48]
3.01 [48]
4.16 [48]
1.05 [48]
0.22 [48]
2.49 [48]
0.76 [48]
4.04 [48]
7.51 [48]
16.71 [48]
7.98 [48]
0.71 [48]
1.11 [48]
1.31 [48]
0.91 [48]
7.79 [48]
7.96 [48]
0.88 [48]
3.28 [48]
1.03 [48]
0.57 [48]
0.41 [48]
1.29 [73]
3.91 [29]
5.08 [29]
1.63 [29]
6.04 [29]
1.04 [29]
2.95 [29]
4.49 [29]
6.72 [29]
1.71 [29]
9.46 [29]
3.7 [181]
3.44 [181]
0.21 [29]
0.24 [29]
2.17 [181]
15.0 [25]
0.82 [118]
0.058 [118]
9.7 [25]
0.046 [181]
13.2 [118]
0.09 [118]
0.096 [118]

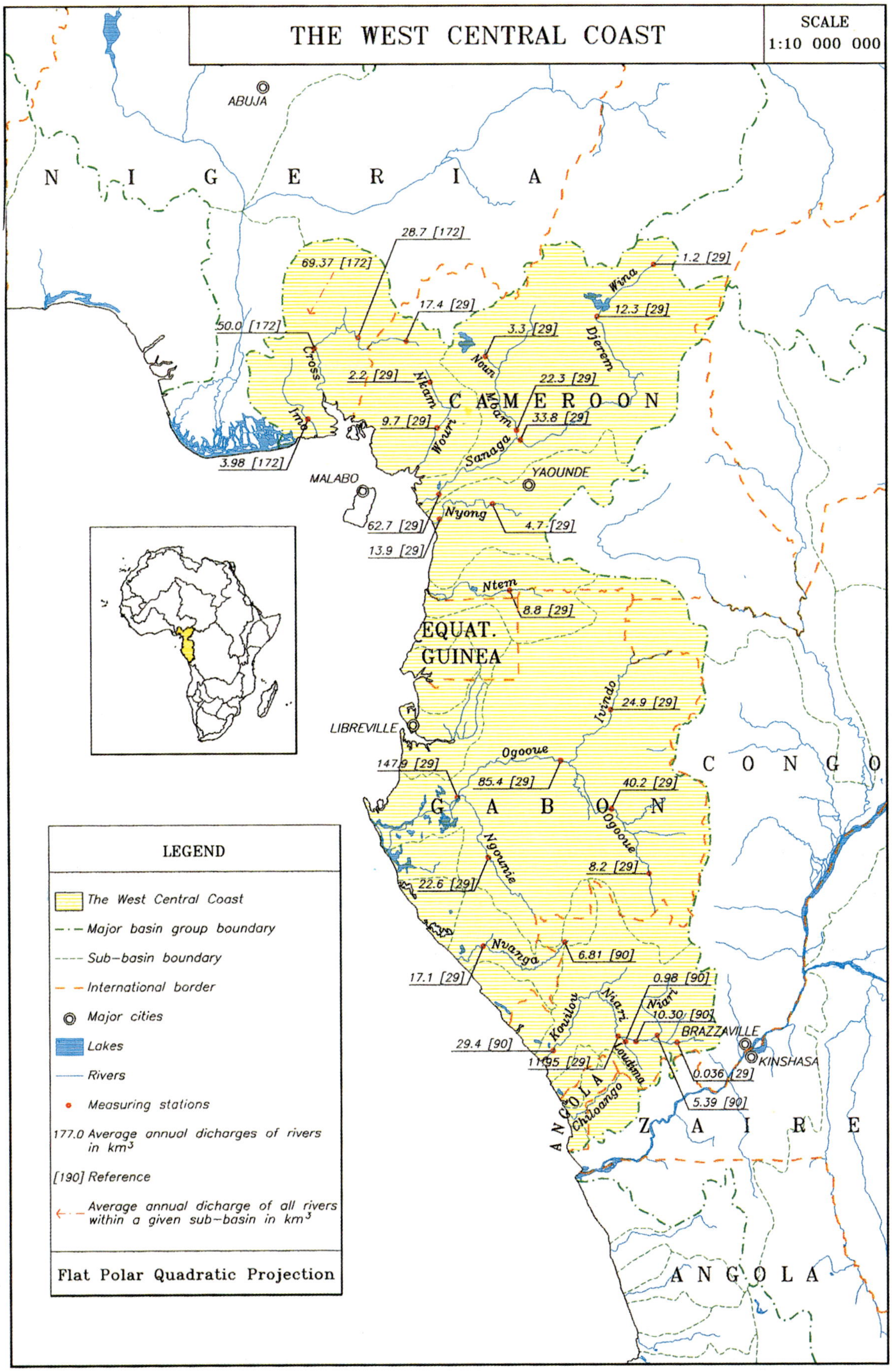
THE WEST CENTRAL COAST
SCALE 1:10 000 000
ABUJA
NIGERIA
CAMEROON
YAOUNDE
MALABO
EQUAT. GUINEA
LIBREVILLE
GABON
CONGO
BRAZZAVILLE
KINSHASA
ZAIRE
ANGOLA
Cross
Ima
Nkam
Wouri
Noun
Mbam
Sanaga
Djerem
Wina
Nyong
Ntem
Ivindo
Ogooue
Ngounie
Nyanga
Kouilou
Niari
Loutima
Chiloango
28.7 [172]
69.37 [172]
17.4 [29]
50.0 [172]
3.3 [29]
1.2 [29]
12.3 [29]
2.2 [29]
22.3 [29]
9.7 [29]
33.8 [29]
3.98 [172]
62.7 [29]
4.7 [29]
13.9 [29]
8.8 [29]
24.9 [29]
147.9 [29]
85.4 [29]
40.2 [29]
8.2 [29]
22.6 [29]
6.81 [90]
17.1 [29]
0.98 [90]
10.30 [90]
29.4 [90]
11.95 [29]
0.036 [29]
5.39 [90]
LEGEND
The West Central Coast
Major basin group boundary
Sub-basin boundary
International border
Major cities
Lakes
Rivers
Measuring stations
177.0 Average annual dicharges of rivers in km³
[190] Reference
Average annual dicharge of all rivers within a given sub-basin in km³
Flat Polar Quadratic Projection

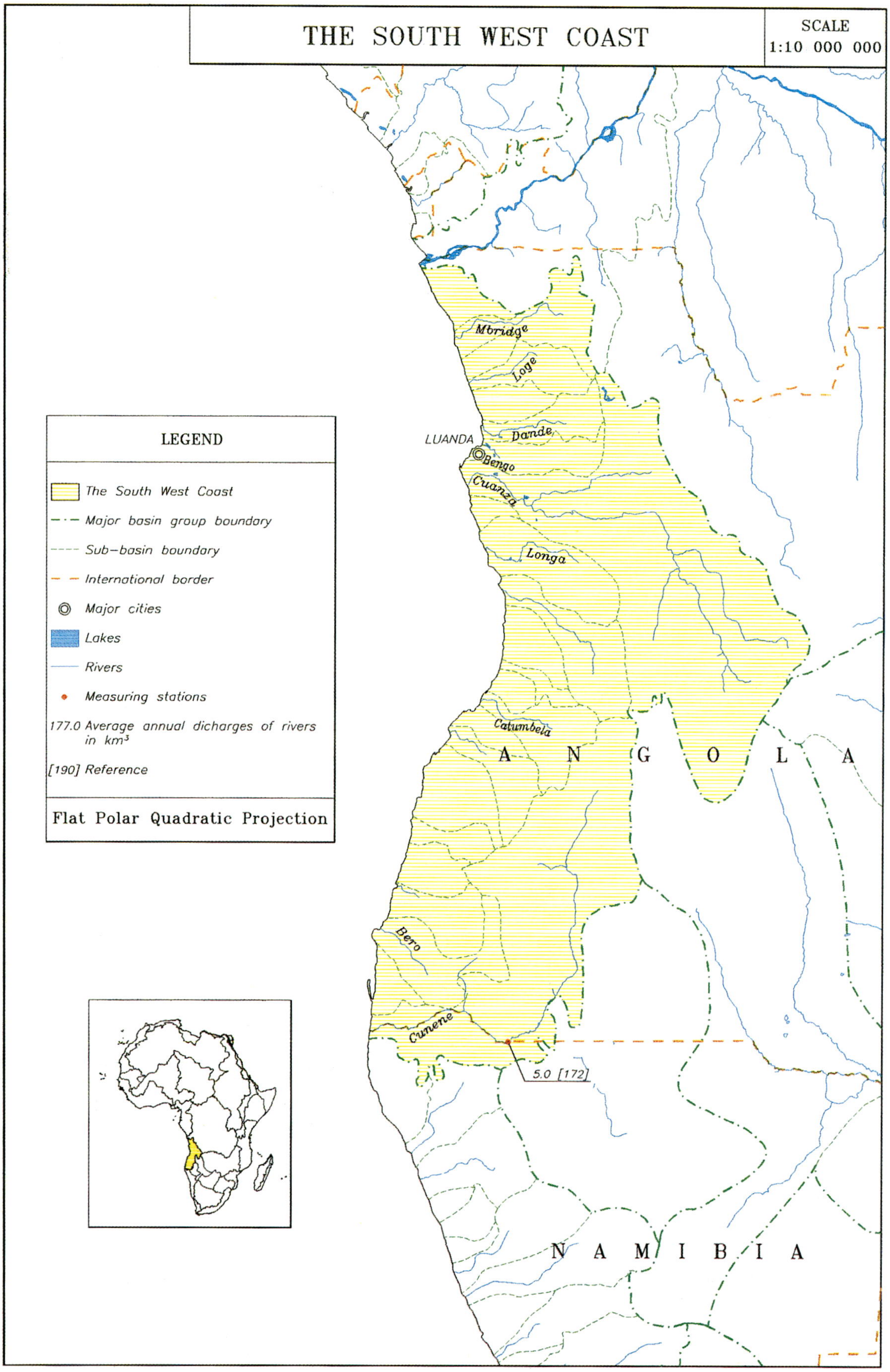
THE SOUTH WEST COAST
SCALE
1:10 000 000
LEGEND
The South West Coast
Major basin group boundary
Sub–basin boundary
International border
Major cities
Lakes
Rivers
Measuring stations
177.0 Average annual dicharges of rivers in km³
[190] Reference
Flat Polar Quadratic Projection
LUANDA
Mbridge
Loge
Dande
Bengo
Cuanza
Longa
Catumbela
Bero
Cunene
5.0 [172]
A N G O L A
N A M I B I A

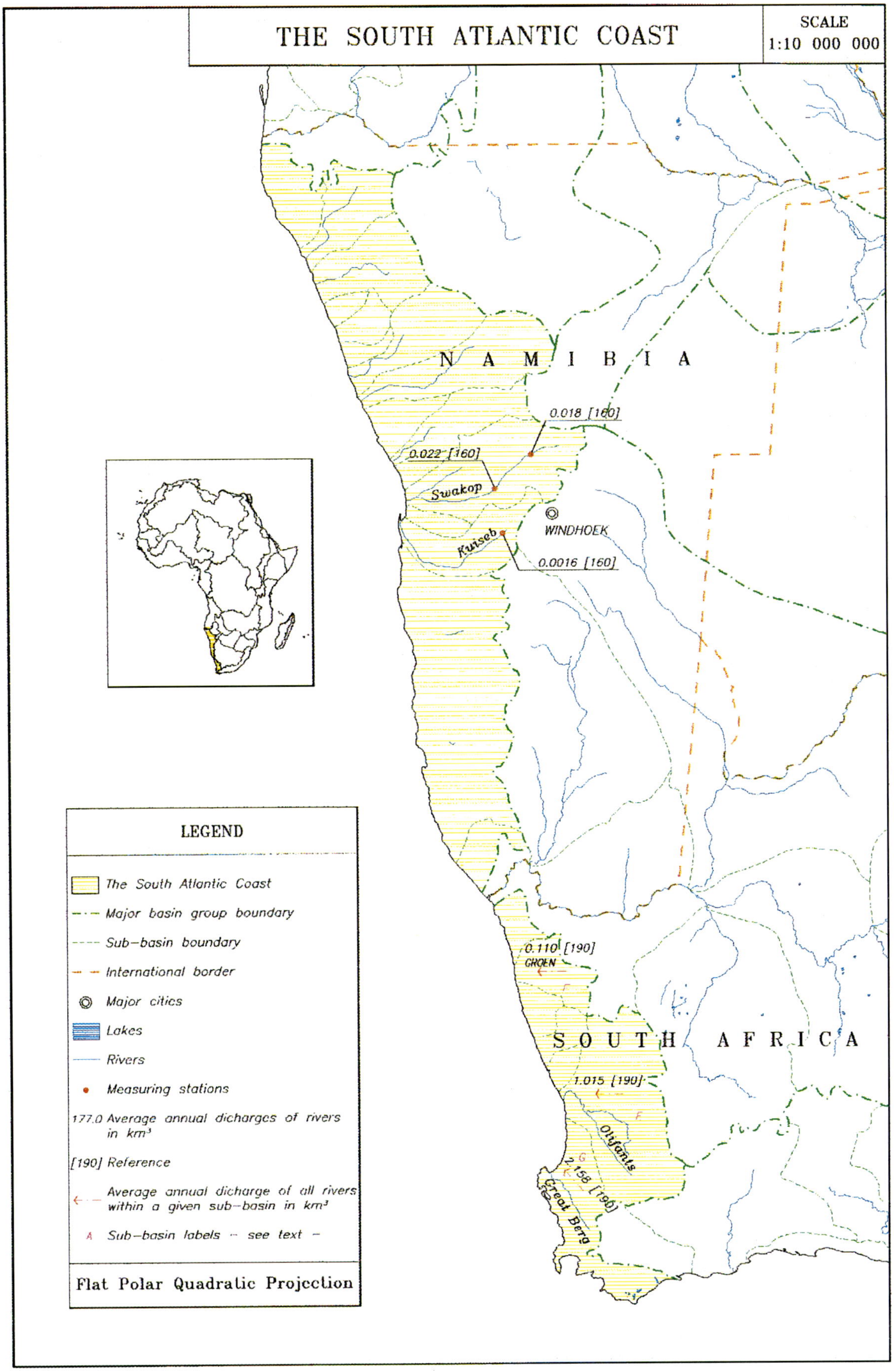
THE SOUTH ATLANTIC COAST
SCALE
1:10 000 000
N A M I B I A
0.018 [160]
0.022 [160]
Swakop
WINDHOEK
Kuiseb
0.0016 [160]
0.110 [190]
GROEN
S O U T H A F R I C A
1.015 [190]
Olifants
2.158 [190]
Great Berg
LEGEND
The South Atlantic Coast
Major basin group boundary
Sub-basin boundary
International border
Major cities
Lakes
Rivers
Measuring stations
177.0 Average annual dicharges of rivers in km³
[190] Reference
Average annual dicharge of all rivers within a given sub-basin in km³
A Sub-basin labels - see text -
Flat Polar Quadratic Projection

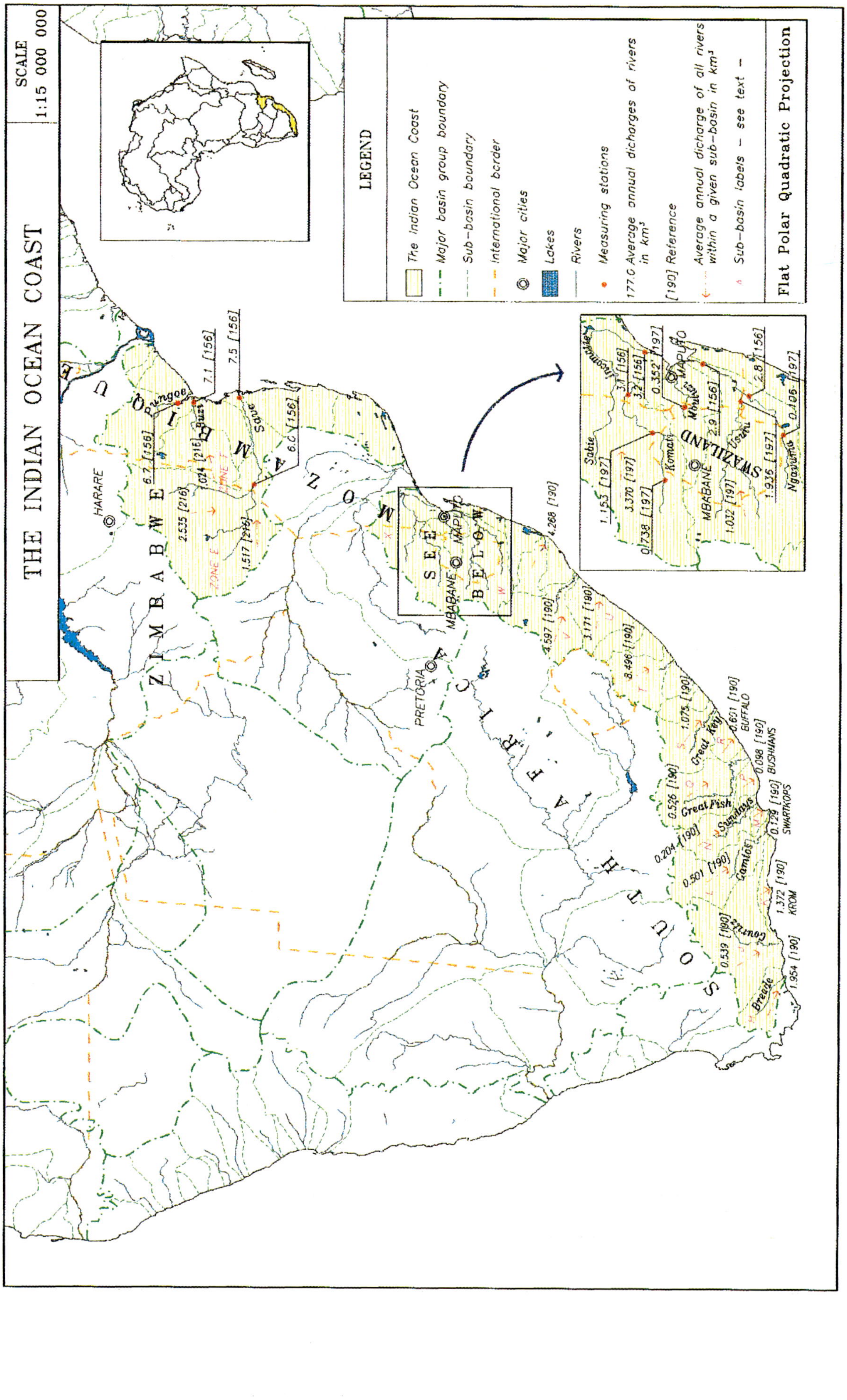
THE INDIAN OCEAN COAST
SCALE
1:15 000 000
LEGEND
The Indian Ocean Coast
Major basin group boundary
Sub-basin boundary
International border
Major cities
Lakes
Rivers
Measuring stations
177.0 Average annual dicharges of rivers in km³
[190] Reference
Average annual dicharge of all rivers within a given sub-basin in km³
Sub-basin labels - see text -
Flat Polar Quadratic Projection
ZIMBABWE
MOZAMBIQUE
SOUTH AFRICA
SWAZILAND
HARARE
PRETORIA
MBABANE
MAPUTO
SEE BELOW
Pungoe
Buzi
Save
Great Fish
Great Kei
Sundays
Gamtoos
Breede
Sabie
Komati
Usutu
Ngwavuma
BUFFALO
BUSHMANS
SWARTKOPS
KROM

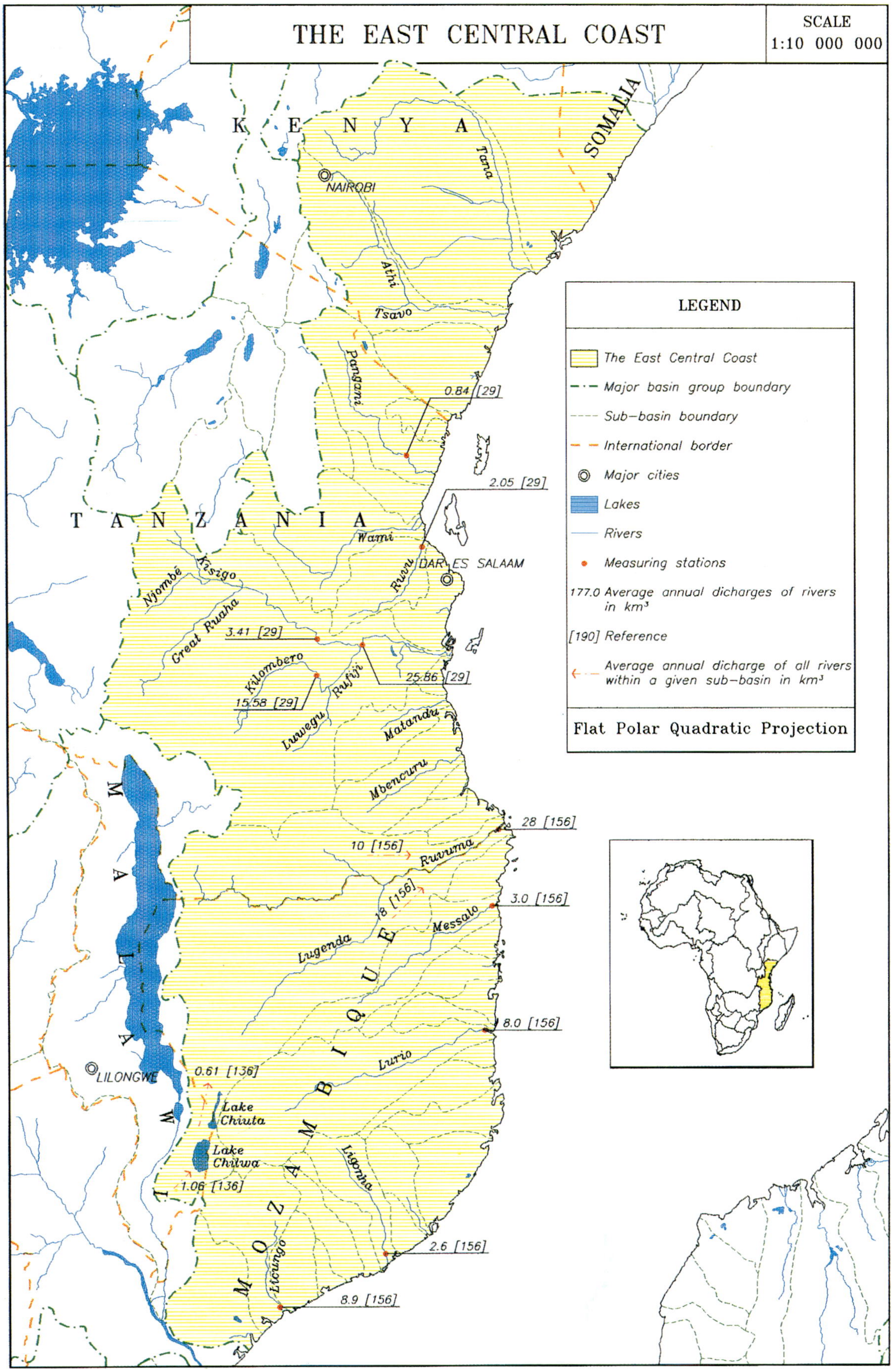
THE EAST CENTRAL COAST
SCALE
1:10 000 000
LEGEND
The East Central Coast
Major basin group boundary
Sub–basin boundary
International border
Major cities
Lakes
Rivers
Measuring stations
177.0 Average annual dicharges of rivers in km³
[190] Reference
Average annual dicharge of all rivers within a given sub–basin in km³
Flat Polar Quadratic Projection
KENYA
SOMALIA
TANZANIA
MALAWI
MOZAMBIQUE
NAIROBI
DAR ES SALAAM
LILONGWE
Tana
Athi
Tsavo
Pangani
Wami
Ruvu
Njombe
Kisigo
Great Ruaha
Kilombero
Luwegu
Rufiji
Matandu
Mbenkuru
Ruvuma
Lugenda
Messalo
Lurio
Ligonha
Licungo
Lake Chiuta
Lake Chilwa
0.84 [29]
2.05 [29]
3.41 [29]
25.86 [29]
15.58 [29]
28 [156]
10 [156]
18 [156]
3.0 [156]
8.0 [156]
0.61 [136]
1.06 [136]
2.6 [156]
8.9 [156]

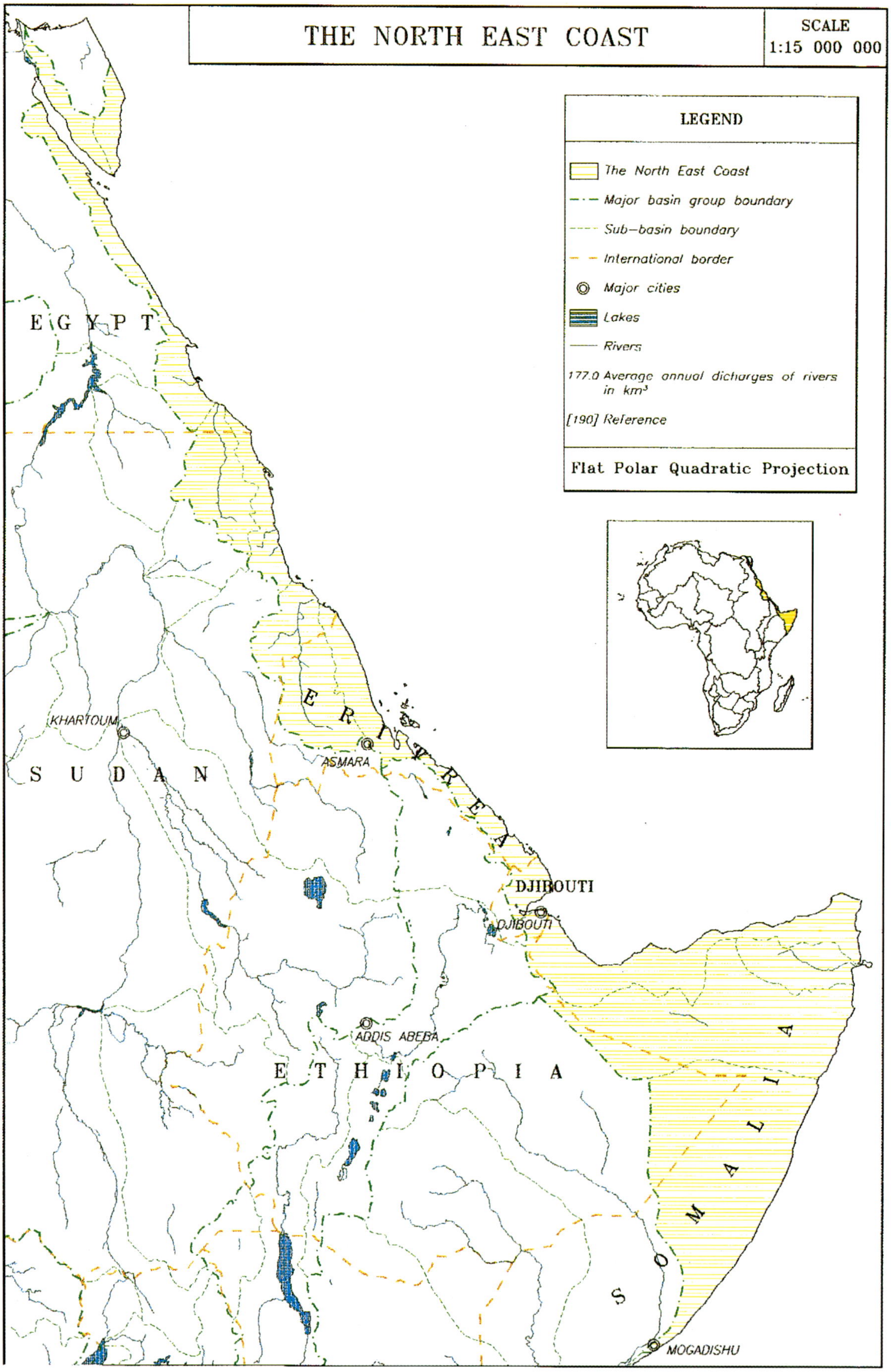
THE NORTH EAST COAST
SCALE
1:15 000 000
LEGEND
The North East Coast
Major basin group boundary
Sub-basin boundary
International border
Major cities
Lakes
Rivers
177.0 Average annual dicharges of rivers in km³
[190] Reference
Flat Polar Quadratic Projection
EGYPT
SUDAN
ERITREA
DJIBOUTI
ETHIOPIA
SOMALIA
KHARTOUM
ASMARA
DJIBOUTI
ADDIS ABEBA
MOGADISHU

Chapter 7

Environmental considerations in irrigation development

Irrigation has contributed significantly to poverty alleviation, food security, and improving the quality of life for rural populations. However, the sustainability of irrigated agriculture is being questioned, both economically and environmentally. The increased dependence on irrigation has not been without its negative environmental effects.

Inadequate attention to factors other than the technical engineering and projected economic implications of large-scale irrigation or drainage schemes in Africa has all too frequently led to great difficulties. Decisions to embark on these costly projects have often been made in the absence of sound objective assessments of their environmental and social implications. Major capital intensive water engineering schemes have been proposed without a proper evaluation of their environmental impact and without realistic assessments of the true costs and benefits that are likely to result.

The sustainability of irrigation projects depends on the taking into consideration of environmental effects as well as on the availability of funds for the maintenance of the implemented schemes. Negative environmental impacts could have a serious effect on the investments in the irrigation sector. Adequate maintenance funds should be provided to the implementing organizations to carry out both regular and emergency maintenance.

It is essential that irrigation projects be planned and managed in the context of overall river basin and regional development plans, including both the upland catchment areas and the catchment areas downstream.

This chapter reviews the most important environmental issues related to irrigation and drainage development.

POTENTIAL ENVIRONMENTAL IMPACTS OF IRRIGATION DEVELOPMENT

The expansion and intensification of agriculture made possible by irrigation has the potential for causing: increased erosion; pollution of surface water and groundwater from agricultural biocides; deterioration of water quality; increased nutrient levels in the irrigation and drainage water resulting in algal blooms, proliferation of aquatic weeds and eutrophication in irrigation canals and downstream waterways. Poor water quality below an irrigation project may render the water unfit for other users, harm aquatic species and, because of high nutrient content, result in aquatic weed growth that obstructs waterways and has health, navigation and ecological consequences. Elimination of dry season die-back and the creation of a more humid micro-climate may result in an increase of agricultural pests an plant diseases.

Large irrigation projects which impound or divert river water have the potential to cause major environmental disturbances, resulting from changes in the hydrology and limnology of river basins. Reducing the river flow changes flood plain land use and ecology and can cause salt water intrusion in the river and into the groundwater of adjacent lands. Diversion of water through irrigation further reduces the water supply for downstream users, including municipalities, industries and agriculture. A reduction in river base flow also decreases the dilution of municipal and industrial wastes added downstream, posing pollution and health hazards.

The potential direct negative environmental impacts of the use of groundwater for irrigation arise from over-extraction (withdrawing water in excess of the recharge rate). This can result in the lowering of the water table, land subsidence, decreased water quality and saltwater intrusion in coastal areas.

Upstream land uses affect the quality of water entering the irrigation area, particularly the sediment content (for example from agriculture-induced erosion) and chemical composition (for example from agricultural and industrial pollutants). Use of river water with a large sediment load may result in canal clogging.

The potential negative environmental impacts of most large irrigation projects described more in detail below include: waterlogging and salinization of soils, increased incidence of water-borne and water-related diseases, possible negative impacts of dams and reservoirs, problems of resettlement or changes in the lifestyle of local populations.

Waterlogging and salinization

About 2 to 3 million ha are going out of production worldwide each year due to salinity problems. On irrigated land salinization is the major cause of land being lost to production and is one of the most prolific adverse environmental impacts associated with irrigation. However, very limited research has yet been conducted to quantify the economic impact of irrigation-induced salinization. Quantitative measurements have generally been limited to the amount of land affected or abandoned. Estimates of the area affected have ranged from 10 to 48% of worldwide total irrigated area. Especially the arid and semi-arid areas have extensive salinity problems.

Waterlogging and salinization of soils are common problems associated with surface irrigation. Waterlogging results primarily from inadequate drainage and over-irrigation and, to a lesser extent, from seepage from canals and ditches. Waterlogging concentrates salts, drawn up from lower in the soil profile, in the plants' rooting zone. Alkalization, the build-up of sodium in soils, is a particularly detrimental form of salinization which is difficult to rectify.

Irrigation-induced salinity can arise as a result of the use of any irrigation water, irrigation of saline soils, and rising levels of saline groundwater combined with inadequate leaching. When surface water or groundwater containing mineral salts is used for irrigating crops, salts are carried out into the root zone. In the process of evapotranspiration, the salt is left behind in the soil, since the amount taken up by plants and removed at harvest is quite negligible. The more arid the region, the larger is the quantity of irrigation water and, consequently, the salts applied, and the smaller is the quantity of rainfall that is available to leach away the accumulating salts.

Excess salinity within the root zone reduces plant growth due to increasing energy that the plant must expend to acquire water from the soil. The tolerance of crops to salinity is variable:

clover and rice are more sensitive to salts than barley and wheat. Comprehensive studies of farm-level effects of irrigation-induced salinity indicate that the yields of paddy and wheat are around 50% lower on the degraded soils and net incomes in salt-affected lands are around 85% lower than the unaffected land.

Irrigation-related salinity has adverse effects not only on the production areas, but also on areas and people downstream. The rivers, particularly in arid zones tend to become progressively more saline from their headwaters to their mouths. The aquifers interrelated with the river are highly saline and the salts discharged to the river system from saline aquifers adversely affect downstream water users, particularly irrigated agriculture and, in some special cases, wildlife.

Many of the soil-related problems could be minimised by installing adequate drainage systems. In Egypt, for example, the installation of drainage systems effectively reduced soil salinity. The average yield for wheat increased from 1 ton/ha before drainage to about 2.4 tons/ha. Similarly, the yield for maize increased from 2.4 tons/ha to 3.6 tons/ha after drainage infrastructure was completed. Drainage is a critical element of irrigation projects, that however still too often is poorly planned and managed. Waterlogging can also be reduced or minimized, in some cases, by using micro-irrigation which applies water more precisely and can more easily limit quantities to no more than the crops needs.

Water-borne and water-related diseases

Water-borne or water-related diseases are commonly associated with the introduction of irrigation. The diseases most directly linked with irrigation are malaria, bilharzia (schistosomiasis) and river blindness (onchocerciasis), whose vectors proliferate in the irrigation waters. Other irrigation-related health risks include those associated with increased use of agrochemicals, deterioration of water quality, and increased population pressure in the area. The reuse of wastewater for irrigation has the potential, depending on the extent of treatment, of transmitting communicable diseases. The population groups at risk include agricultural workers, consumers of crops and meat from the wastewater-irrigated fields, and people living nearby. Sprinkler irrigation poses an additional risk through the potential dispersal of pathogens through the air.

The risk that one or more of the above diseases is introduced or has an increased impact is most likely in irrigation schemes where [8]:

- soil drainage is poor, drainage canals are either absent, badly designed and/or maintained;
- rice or sugar cane is cultivated;
- night storage reservoirs are constructed;
- borrow pits are left with stagnant water;
- canals are unlined and have unchecked vegetation growth.

Malaria

Malaria is by far the most important disease, both in terms of the number of people annually infected, and whose quality of life and working capacity are reduced, and in terms of deaths. Worldwide, some 2 000 million people live in areas where they are at risk of contracting malaria. The total number of people infected with malaria is estimated at 100 to 200 million with between 1 and 2 million deaths per year, with almost 90% of the cases in Africa. Drug treatment has become difficult recently because the parasite has become resistant to certain drugs that have been used for a long time in many parts of the world. Interruption of disease transmission using chemicals for the control of the vector, the mosquito, has become less effective because some mosquito vector species have become resistant to previously effective insecticides and some insecticides have been banned for environmental reasons.

Bilharzia

Bilharzia is almost as widespread as malaria, but rarely causes immediate death. An estimated 200 million people are infected and the transmission occurs in some 74 countries. The infection is particularly common in children who play in water inhabited by the snail intermediate host. Severe infection in childhood leads to long-term damage to bladder, kidneys and liver, which may cause death many years after the original infection. Infection at any age may make people feel unwell and reduce working capacity.

Bilharzia is an infection caused by parasitic worms or blood flukes of certain species of the genus *Schistosoma*. Adult parasites live in the blood of mammals, but their life cycle requires a phase of asexual multiplication within a fresh-water snail host. The flukes infect humans who enter their exposed skin in water, usually through swimming, bathing or wading. There exists either urinary or intestinal schistosomiasis. The type and extent of health complications associated with schistosomiasis appear to vary with species and strain of parasite and by the characteristics of the human population.

As shown in some examples below, water resources development projects may make schistosomiasis worse, and that there are serious public health consequences in many cases. Specifically, WHO stated that in areas endemic for schistosomiasis, water resources development projects should have schistosomiasis prevention and control built into programme design and implementation. Furthermore, even in cases where irrigation does not increase schistosomiasis infection rates, careful studies should be made of snail species and of existing patterns of schistosomiasis transmission. Further, a percentage of investment and operating funds should be allocated for appropriate water supply and sanitation and for health care to treat local populations for any water-related or other ailments associated with the project.

Effects of large dams and reservoirs on the prevalence of bilharzia in Africa

In Africa, in recent decades cautionary warnings have accompanied many irrigation and dam projects regarding the likely impact of increased schistosomiasis transmission. In some cases, these prophecies came true, including the Volta Lake project in Ghana in the 1960s and the Kainji Lake project in Nigeria around 1970. Most recently, research appears to confirm the prediction, that constructing a dam across the mouth of the Senegal River would lead to a surge in schistosomiasis transmission.

In part of Upper Egypt, schistosomiasis prevalence is said to have increased from 6 to 60% three years after the Aswan low dam was completed in the early 1930s. Following the construction of the Sennar dam in Sudan in 1926 and the Gezira scheme which followed, schistosomiasis spread. At the Arusha Chini irrigation project in Tanzania, research reported the prevalence to be 85% in 1962. On the southwestern shore of Lake Volta in Ghana, prevalence is reported to have reached 90% two years after the lake was filled in 1966.

In Zambia, prevalence of schistosomiasis around Lake Kariba in 1968, 10 years after the Zambezi river was dammed, was 15% in adults and 70% in children. At Kainji Lake in Nigeria, prevalence increased from 30% to 45%.

In Ethiopia, the two Koka dams in the Awash Valley present a case of the different effects a dam has on the two different forms of schistosomiasis through the reservoir, downstream hydrology, and irrigation. It was reported that these two dams, built in 1958 and 1964, appeared to have no effect on urinary schistosomiasis transmission through their reservoirs as assessed in 1968. The dams did, however, enable intestinal schistosomiasis transmission to occur in the upper Awash Valley by changing its hydrology and introducing irrigation.

In West Africa, both urinary and intestinal forms of schistosomiasis became highly endemic in the Office du Niger area. The irrigation scheme had a mean prevalence of urinary schistosomiasis of 64.4%, compared to the river communities' 19.9%, and a mean prevalence of intestinal schistosomiasis of 53.9%, compared to the river communities' 1.9%.

Effects of small dams on the prevalence of bilharzia in Africa

Throughout the semi-arid areas of Africa, people have constructed many small earthen dams to provide irrigation water for dry season cultivation. While reports often blame these dams for spreading schistosomiasis, there is little evidence to substantiate such claims.

Small dams in the semi-arid zone of West Africa get the greatest amount of attention. Although the dams built for dry-season irrigation extended the range of *Bulinus rohlfsi* snails, a common vector for urinary schistosomiasis in West Africa, this did not cause any noticeable increases in prevalence of urinary schistosomiasis.

Several studies in the same Sudano-Sahel ecological zone as northern Nigeria noted that evidence linking small earthen dams to schistosomiasis was lacking. In Burkina Faso, natural seasonal ponds infested by *Bulinus truncatus snails* are more common and dangerous locations of infections than are artificial reservoirs created by small dams.

In Cameroon, a nation-wide survey failed to find an association between the rate of schistosomiasis and small dams. Instead, temporary ponds and snail hosts adapted to low seasonal rainfall permits intense transmission of urinary and intestinal schistosomiasis throughout northern Cameroon, regardless of dams.

Contrary to the experience in Nigeria, Burkina Faso and Cameroon, one study in northern Ghana showed small dams to be linked to schistosomiasis. Data collected during 1960-61 showed much higher prevalence of schistosomiasis in the eastern, more densely settled part of what was the Upper region of Ghana. The mean prevalence of urinary schistosomiasis in the region's eastern part was 19.8% in 15 districts without dams, 42.3% in 16 districts with dams 1-2 years old, and 52.0% for 6 districts with dams 3 years old. Areas in the western part of the region with few or no dams had infection rates under 10% in 6 districts, and 10 to 29% in 10 others. In contrast, prevalence was over 70% in 2 of the 3 western districts containing dams.

The control of the water-related diseases

The control of the water-related diseases can be effected in a number of ways, some of which are mutually reinforcing. Three types of measures are distinguished [8]:

- measures aimed at the pathogens: immunization, prophylactic or curative drugs;
- measures aimed at reducing vector densities or vector lifespan: chemical, biological and environmental controls;
- measures to reduce human/vector or human/pathogen contact: health education, personal protection measures and mosquito proofing of houses.

Of the above, environmental control measures are considered to be long-lasting and environmentally-sound. These include preventing or removing aquatic vegetation, lining canals with cement or plastic, regularly fluctuating water levels, periodic rapid drying of irrigation canals, preventing contamination of water bodies with faeces, supply of safe and clean drinking water, appropriate siting of housing of the farmers etc. For example, in Zimbabwe, in a communal small-holder irrigation project at Mushandike, adoption of a these measures resulted within three years in a drop of the infection rate from an initial 70 to 80% to virtually nil.

Potential environmental impacts of dams and reservoirs

The benefits of a dam project are flood control and the provision of a more reliable and higher quality water supply for irrigation, domestic and industrial use. Intensification of agriculture locally through irrigation can reduce pressure on uncleared forest lands, intact wildlife habitat and marginal agricultural land. In addition, dams create reservoir fishery and the possibilities for agricultural production on the reservoir drawdown area, which more than compensate for losses in these sectors due to the dam construction.

However, large dam projects cause irreversible environmental changes over a wide geographic area and thus have the potential for significant impacts. Criticism of such projects has grown in the last decade. Severe critics claim that because benefits from dams are outweighed by their social, environmental and economic costs, the construction of large dams is unjustifiable. In some cases, environmental and social costs can be avoided or reduced to an acceptable level by carefully assessing potential problems and implementing cost-effective corrective measures.

Damming the river and creating a lake-like environment profoundly changes the hydrology and limnology of the river system. Dramatic changes occur in the timing of flow, quality, quantity and use of water, aquatic biota, and sedimentation in the river basin. The area of influence of a dam project extends from the upper limits of the catchment of the reservoir to as far downstream as the estuary, coast and offshore zone. While there are direct environmental impacts associated with the construction of the dam (for example dust, erosion, borrow and disposal problems), the greatest impacts result from the impoundment of water, flooding of land to form the reservoir and alteration of water flow downstream. These effects have direct impacts on soils, vegetation, wildlife and wildlands, fisheries, climate and especially the human populations in the area.

Increased pressure on upland areas above the dam is a common phenomenon caused by the resettlement of people from the inundated areas and by the uncontrolled influx of newcomers into the basin catchment. On-site environmental deterioration as well as a decrease in water quality and increase in sedimentation rates in the reservoir result from clearing of forest land for agriculture, grazing pressures, use of agricultural chemicals, and tree cutting for timber or fuelwood.

The impact of dams on flooding

The function of dams and reservoirs in flood control is to reduce the peak flows entering a flood prone area. Rather than maintaining high water levels for increased head or sustained water supply for irrigation, flood control operation requires that water levels be kept drawn down deliberately prior to and during the flood season in order to maintain the capacity to store any incoming floodwater. However, flood plains may be productive environments because flooding makes them so. Flooding recharges soil moisture and replenishes the rich alluvial soils with flood deposits of silt. In arid areas flooding may be the only source of natural irrigation and soil enrichment. Reduction or elimination of flooding has the potential for impoverishing flood recession cropping, groundwater recharge, natural vegetation, wildlife and livestock population in the flood plain which are adapted to the natural flood cycles.

To maintain the productivity level of the natural systems, compensatory measures have to be taken, such as fertilization or irrigation of agricultural lands. In addition, when channelization measures reduce the frequency of flooding, the sediments entering the river systems from catchment areas upstream will be passed to the mouth of the river unless overflow areas are present downstream. Channel modification can result in a number of negative environmental impacts. Any measure that increases the velocity of flow increases the erosive capacity of the water. Although channel improvement can alleviate flooding problems in the treatment area, flood peaks are likely to increase downstream, thus simply transferring the problem elsewhere. Dikes built on the flood plain to exclude water from certain areas affect the hydrology of the area, and can have impacts on wildlife and livestock habitat and movement.

The impact of dams on fisheries and wildlife

Fishery alongside the rivers usually declines due to changes in river flow, deterioration of water quality, water temperature changes, loss of spawning grounds and barriers to fish migration. A reservoir fishery, sometimes more productive than the previous fishery alongside the river, however, is created.

In rivers with biologically productive estuaries, both marine and estuarine fish and shellfish suffer from changes in water flow and quality. Changes in freshwater flows and thus the salinity balance in an estuary will alter species distribution and breeding patterns of fish. Changes in nutrient levels and a decrease in the quality of the river water can also have profound impacts on the productivity of an estuary. These changes can also have major effects on marine species which feed or spend part of their life cycle in the estuary, or are influenced by water quality changes in the coastal areas.

The greatest impact on wildlife will come from loss of habitat resulting from reservoir filling and land use changes in the catchment area. Migratory patterns of wildlife may be disrupted by the reservoir and associated developments. Aquatic fauna, including waterfowl, amphibians and reptiles can increase because of the reservoir.

Socio-economic impacts of irrigation schemes

The objective of irrigation projects is to increase agricultural production and consequently to improve the economic and social well-being of the rural population. However, changing land use patterns may have other impacts on social and economic structure of the project area. Small plots, communal land use rights, and conflicting traditional and legal land rights all create difficulties when land is converted to irrigated agriculture. Land tenure/ownership patterns are almost certain to be disrupted by major rehabilitation works as well as a new irrigation project. Similar problems arise as a result of changes to rights to water. Increased inequity in opportunity often results from changing land use or water use patterns. For example, owners benefit in a greater proportion than tenants or those with communal rights to land. Access improvements and changes to the infrastructure are likely to require some field layout changes and a loss of some cultivated land.

Irrigation projects tend to encourage population densities to increase, either because of the increased production of the area or because they are part of a resettlement project. Impacts resulting from changes to the demographic/ethnic composition may be important and have to be considered at the project planning stage through, for example, sufficient infrastructure provision.

The most significant issue arising from large dam construction is resettlement of people displaced by the flooding of land and homes. This can be particularly disruptive to communities and insensitive project development would cause unnecessary problems by lack of inadequate compensation of the affected population. Human migration and displacement are commensurate with a breakdown in community infrastructure which results in a degree of social unrest and may contribute to malnutrition. As an example of the number of people displaced by the construction of a dam, filling of the reservoir behind the High Aswan Dam displaced 50 000 to 60 000 people in Egypt and some 53 000 people in the Sudanese portion.

Changing land patterns and work loads resulting from the introduction or formalizing of irrigation are likely to affect men and women, ethnic groups and social classes unequally. Groups that use common land to make their living or fulfill their household duties, for example for charcoal making, hunting, grazing, collecting fuel wood, growing vegetables, etc. may be disadvantaged if that same land is taken over for irrigated agriculture or for building irrigation infrastructure. Women, migrants groups and poorer social classes have often lost access to resources and gained increased work loads. Conversely, the increased income and improved nutrition from irrigated agriculture may benefit women and children in particular.

The most common socio-economic problems reducing the income generating capacity of irrigation schemes are:

- The social organization of irrigation operation and maintenance (O&M). Poor O&M contributes significantly to long-term salinity and waterlogging problems and needs to be adequately planned at the design stage to sustain the long-term development of the schemes.
- Reduced farming flexibility. Irrigation may only be viable with high-value crops, thus reducing extensive activities such as grazing animals, operating woodlots, etc.
- Changing labour patterns that make labour-intensive irrigation unattractive.
- Insufficient external supports such as markets, agrochemical inputs, extension and credit facilities.

User participation at the planning and design stages of both new schemes and the rehabilitation of existing schemes, as well as the provision of extension, marketing and credit services, can minimize negative impacts and maximize positive ones.

Alternatives to mitigate the negative impacts of irrigation projects

Alternatives exist to mitigate adverse effects of irrigation development. Some of them are listed below:

- locating the irrigation project on the site where negative impacts are minimized;
- improving the efficiency of existing projects and restoring degraded croplands to use rather than establishing a new irrigation project;
- developing small-scale, individually-owned irrigation systems as an alternative to large-scale, publicly-owned and managed schemes;
- using sprinkler irrigation and micro-irrigation systems to decrease the risk of waterlogging, erosion and inefficient water use;
- using treated wastewater, where appropriate, to make more water available to other users;
- maintaining flood flows downstream of the dams to ensure that an adequate area is flooded each year, among other reasons, for fishery activities.

THE ROLE OF WETLAND AND THE IMPACTS OF WATER DEVELOPMENT PROJECTS

Wetlands are wildlands of particular importance both economically and environmentally. The most important roles which wetlands perform are:

- Preservation of biological diversity: for many species of shrimp, fish and waterfowl, tidal and fresh water marshes, coastal lagoons and estuaries are of vital importance as breeding grounds as well as staging areas in their migration routes. All types of wetlands may harbour unique plants and animals.

- Production of goods: wetlands are among the most productive ecosystems in the world. Estuaries and tidal wetlands, in particular mangroves, are important nursery areas for most species of fish and shrimp which are later caught offshore. Shallow water areas are, in general, rich fishing grounds. Flood plains are important grazing areas for cattle and wildlife and vital spawning grounds for many fish species. Swamp forest may yield valuable timber.

- Production of services: wetlands can contribute to local rainfall and can be an efficient, low cost water purification system (herbaceous swamps), a recreation area (hunting, fishing, boating), a buffer against floods, and provide protection against coastal erosion by storms (mangroves).

Despite their importance, wetlands are under threat, in particular, from direct conversion of wetlands for agriculture and projects which affect the hydrology of a wetland, such as construction of dams, flood control, lowering of the aquifer drainage, and irrigation and other water supply systems.

The sections below describe some important wetlands in Africa, but the list is far from exhaustive.

Wetlands in the West African Sahel

The Senegal river waters the west of Mali, the north of Guinea, the north of Senegal and the south of Mauritania (Map 1). the main Niger river stretches through Mali, the Republic of Niger and Nigeria (Map 2). The frontiers of Niger, Nigeria, Cameroon and Chad meet in the Lake Chad (Map 3). For all of these countries, river valleys and lakes are of the utmost economic importance because of their high productivity. Mauritania, Mali, Niger and Chad, whose territories are for the most part desert, draw their agricultural resources and the greater part of their animal protein from fishing and stock-raising in the river valleys. The states along the water courses are acting in concert to increase the use of water resources and avoid the hazards of drought. It is a question of ensuring mastery over the flood waters to grow rice and other cereals, and of using surface water for irrigation. These operations inevitably tend to modify ecological conditions. In practice this will lead to significant losses of aquatic habitats and to a noticeable decrease in fish and waterfowl.

Regions such as the southern part of the Sahel, with a strongly seasonal rainfall regime and yet with sufficient rainfall to support seasonal agriculture and pastoralism, support large numbers of people. Flood plains, swamps and lakes provide a range of ecological resources and economic opportunities. Without wetlands, the drylands of the West African Sahel would be both less productive and more hazardous as a place for people to live.

In the semi-arid zone of Western Africa, patterns of rainfall and river flow are strongly seasonal. In northern Nigeria, most of the rainfall occurs in just three or four months, between June and September. During this short rainy season, precipitation exceeds evapotranspiration and runoff occurs. Savannah rivers run strongly, but start to shrink as the rains end. During the wet season rainfed agriculture is possible, and there are extensive grasslands providing relatively nutritious grazing for livestock away from the river valleys. Once the rains end, these resources also dry up, and pastoralists concentrate on the remaining wetlands in river valleys or larger wetlands such as the delta of the Senegal River, the Niger Inner Delta in Mali or Lake Chad.

It is also only in these areas that agriculture can continue into the dry season. In many areas rice is planted, either in the rains before the floods arrive, or as the floods recede. Flood-recession crops such as sorghum or beans are planted as the waters recede, and farmers dig wells or use remaining pools to irrigate small gardens using buckets, shadoofs or, more recently, small pumps. At the same time, other economic activities such as fishing are also linked to the changing flood. Many fishes move laterally out of the riverbed pools into the flood plain to breed, and their offspring is caught as the water retreats. Later in the dry season, the residual pools in the riverbed are themselves fished. The high fisheries productivity of most of the seasonally inundated flood plains is fostered, at least in part, by the nutrient-rich dung left by the grazing animals during the previous dry season.

In valleys such as the Senegal or the vast flooded plains of the Niger Inner Delta, the annual cycle of farmers, herders and fishers is closely linked to the seasonal cycle of flooding. The flood plain of the Senegal stretches up to 30 km in width, and runs 600 km downstream of Bakel. It covers a total of about 1 million ha and supports farmers, pastoralists and fishing communities. Up to half a million people depend on the flood-related cropping in the '*waalo*' land of the flood plain.

There is growing evidence that large-scale capital intensive water development schemes do provide neither the range of foodstuffs nor the economic return of traditional systems.

Studies, comparing the efficiency per unit of water of traditional extensive systems of cultivation, grazing and fishing in the Niger Inner Delta with the intensive modern irrigation project of the Office du Niger, showed that both systems produce about the same gross profit margin, even when the running costs and management charges for the irrigation scheme are taken into account. However, the extensive system produces meat, milk, fish and rice compared to the rice-only irrigated system. More importantly, when the interest charges arising from the irrigation scheme are taken into account, the net profit from the irrigated rice turns into a loss of \$0.65/100 m^3 of water, whilst the extensive traditional methods benefit from the 'free services of nature' and turn in a net profit of \$0.42/100 m^3 at 1984/85 prices [15a].

The Hadejia-Nguru Wetlands

The Hadejia-Nguru Wetlands [170] concern a part of the flood plain of the Komadougou-Yobe river basin in the Lake Chad basin in the north-east of Nigeria (Map 3) and are home to probably about a million people. The wetlands have formed where the waters of the Hadejia and Jama'are rivers meet the lines of ancient sand dunes aligned northeast-southwest. An area of confused drainage has formed here, with multiple river channels and a complex pattern of permanently and seasonally flooded land and dryland. The wetlands are nationally and internationally important for migratory waterfowl. The wetlands support extensive wet-season rice farming, flood-recession agriculture and dry-season irrigation. The flood plain also supports large numbers of fishing people, most of whom also farm, and is grazed by very substantial numbers of Fulani livestock, particularly cattle, which are brought in from both north and south in the dry season. There is also an important dispatch from the wetlands of fuelwood and fodder for horses. In the past, much of the rice, as well as fish and birds, was traded out of the area. This has changed, but there is now a strong export of other agricultural products, for example peppers, wheat and fuelwood. The economic value of production from the wetlands is very large, many times greater than that of all the irrigation schemes for which the inflowing rivers are dammed, diverted and their waters used.

There are natural changes, for example the impacts of drought, that have serious implications for the future of the wetlands and the sustainability of their production systems. There are also major economic changes within the wetlands themselves. The extent of irrigation has greatly increased over the 1980s, largely as a result of the advent of small petrol-powered pumps and the ban on the importation of wheat in 1988. As the use of small pumps spreads, conflicts are beginning to emerge between farmers and pastoralists, and between small and large farmers for access to land.

The wetlands have also been affected by developments elsewhere in the river basin. The construction of the Tiga Dam on a tributary of the Hadejia river in the early years of the 1970s has exacerbated the effects of the low rainfall of the last two decades. The result has been a reduction in the extent of flooding in the wetland. Construction of a dam on the Hadejia river just above Hadejia town to provide short-term storage of water to irrigate the Hadejia Valley Project Phase 1 began in the early 1980s, but was stopped for several years because of financial problems. The main dam was completed in 1992, soon after work restarted on the project. The dam has created a large shallow lake upstream and it will probably have a major effect on the timing and extent of flooding in the wetlands.

Most of the dams, irrigation schemes and water resources plans for the Yobe basin were prepared in the 1970s and early 1980s, using data for the relatively wet period up to 1973. The post-1972 drought has reduced the proportion of rainfall which runs off to the rivers. The 1988 flood at Hadejia was probably one of the largest for some years and it was augmented by the failure of the dam at Bagauda.

The Hadejia-Nguru wetlands have long been known as a centre of fish production. Upstream hydrological developments induced by irrigation projects threaten to degrade this important resource. Studies of flood plain fisheries have shown that fish production is closely related to flood extent. The existing and planned dams upstream of the Hadejia-Nguru wetlands are likely to have a serious impact on fisheries. Despite the lack of information specific to the Hadejia-Nguru wetlands, there are enough studies from other flood plains affected by hydraulic works to show that the effects of dams on fish communities are likely to be serious. The dams are likely to bring changes in river flow, loss of habitat, blocking of channels, changes in silt loading, plankton abundance and temperature which are likely to affect fish communities.

The economic value of fish production from the flood plains adds weight to the argument in favour of maintaining the annual flooding of the wetlands. Moreover, the significance of fishing goes beyond its value in monetary terms. Fishing plays an important role in the flexibility and adaptability of the rural economy in the flood plains. A reduction in this flexibility through degradation of the fishery resource may have serious repercussions on the ability of communities to adapt to fluctuations in their environment. Many people are involved in the fisheries and so the social consequences of any appreciable reduction in productivity will be felt throughout the area. Degradation of the fisheries may also affect other sectors of the rural economy. Most people who fish also pursue other activities - such as farming, livestock rearing, manufacturing of crafts or trading - and the loss of, or reduction in one component of the household economy is likely to affect activities in other sectors. There will also be 'downline' effects on fish processors, fish dealers, customers and consumers.

In addition to producing fuelwood, the forest reserves and bushland of the flood plains yield important non-timber forest products that are significant to the livelihoods and subsistence of local communities. Some, including leaves, are important marketed commodities that generate substantial income. *Doum* palm leaves are either processed into mats and other products or sold as raw material. The harvesting and processing of doum palm leaves is a dry season activity, and many people migrate to the wetlands to harvest the palm. Mat-making from doum is also a specialized activity of many flood plain villages. Mats and other doum products, for example rope and baskets, are sold locally or exported to other regions. *Baobab* leaves are used widely as an ingredient for soups and stews and are especially important as a 'drought food'. Honey, produced by local beekeepers, is a highly valued commodity.

Since 1985, the area has been the focus of the Hadejia-Nguru Wetlands Conservation Project. This project has been run jointly by the Nigerian Conservation Foundation, IUCN (International Union for the Conservation of Nature), the Royal Society for the Protection of Birds and the International Council for Bird Preservation (now renamed Birdlife International). In 1990 a major development project was started by the European Community that included the eastern part of the area. The North East Arid Zone Development Project (NEAZDP) has a very substantial budget to generate village-based development initiatives. Attention has tended to be directed in particular to the potential resources of the wetlands.

Wise use of the wetlands of the Hadejia-Nguru wetlands demands a proper understanding of the environmental and socio-economic changes that are occurring and of those that may be predicted. Understanding of the impacts of changes inside and outside the flood plain is far from easy, and prediction of future impacts is even harder. However, without such understanding and prediction, effective planning and management is impossible.

The economic importance of the flood plains suggests that benefits it provides cannot be excluded as an opportunity cost of any scheme that diverts water away from the flood plain system. Policy makers should be aware of this problem when designing water development projects in the river system. Further analysis is also required of the type of 'regulated flood projects' regime, which could maintain much of the flood plain system intact while still allowing some upstream water developments. Further investigation of all the economic benefits provided by the wetlands is also needed, and the sustainability of production within a flood plain area should be more thoroughly examined.

Effects of the Jonglei Canal on the Sudd swamps

In the southern of Sudan, the Nile discharges its water into the great wetlands of the Sudd, a network of channels, lakes and swamps in which as much as half of the inflowing water is disappears through evaporation (see also Chapter 6: *the Nile basin section*). The Jonglei Canal was designed to bypass the Sudd and direct downstream a proportion of the water that is 'lost' from the Nile each year by spill and evaporation in the swamps. The projected dimensions of the canal are as follows: a width from bank to bank of about 75 metres, a channel bed-width averaging 38 metres, a depth varying from 4 to 8 meters, and a length of 360 km, over twice the length of the Suez canal. Jonglei is a small Dinka village close to the Atem channel at a point where the canal alignment was planned to begin. Although the offtake will now be further south at Bor, the canal is still so named and Jonglei has given its name to a province as well.

The canal has not been completed, but detailed surveys were undertaken to determine a whole range of effects, many of which will be shown to be disadvantageous to the inhabitants of the Jonglei Area. Some of the effects are described below.

The river-flooded grasslands are an essential seasonal resource during the driest months of the year. Not only is there drinking water available in the rivers, but the process of seasonal inundation itself produces species of grasses which sustain the herds from about January until April. There are no other alternatives as the grasses of the high land are exhausted or reserved for the livestock (mainly smaller stock), held by the few people who elect to remain behind, and the rain-flooded grasslands have become woody and unpalatable and produce little or no regrowth after burning. It follows that the river-flooded grasslands are crucial to the pastoral economy at this time of the year. It is, however, just these grasslands that may be reduced by the operation of the canal.

The water benefit of the canal downstream will be around 4 km^3/year and according to some estimates even an extra water flow of up to 10 km^3/year may be reached. These quantities are a substantial percentage of the average 'losses' by the evapotranspiration, the natural production of river-flooded grasses being a function of the annual fluctuation in river discharge and thus of the annual variation in area flooded. In other words, to the local inhabitants these are not losses in water at all, though the waters are excessive and the cause of damaging floods, as in 1964.

The floods of the 1960s, reaching a peak in 1964, caused great damage to human interests. On the Zeraf island alone it was reckoned that 130 000 cattle were lost owing to exposure and lack of grazing since practically the whole area remained under water for a long period. Similar disastrous effects occurred west of the Bahr el Jebel in the vicinity of Adok. It follows that any reduction in peak flows could be protective and beneficial. The same model can be applied to give some indication of the effect of the canal on areas of flooding. The figure of 25 million m^3/day for a canal diversion may reduce the area of flooding by about 19% at a 1964 peak discharge [41].

The established fisheries of some large lakes in the Sudd are said to have been adversely affected by increased water depth, but, overall, the flooding of the 1960s has multiplied the number of perennial lakes in the system and, thereby, the fishing potential. A severe decrease in the discharge into the Sudd resulting from the Jonglei canal would bring about the total disappearance of many lakes in the papyrus zone and reduce others to the status of seasonal lagoons, with a serious loss of year-round fish and fishing potential.

The Jonglei Canal brings the obvious advantage of shorter river communications between Khartoum and the main urban centre of the southern Sudan at Juba, in effect reducing the length of the journey by 300 km. The canal will also bring communications, as well as water, to a particularly remote part of Sudan, which is inaccessible during the rains and largely abandoned in the dry season. Passing points and berthing places are part of the design and will lead to the creation of small ports which are likely to develop and open up contact with the hinterland in much the same way as those along the natural channels of the river have done. There is, however, likely to be considerable disadvantage to the people of the Zeraf Island and those living west of the Bahr el Jebel, in that mainstream traffic will follow the canal and the old western landing places will be ill-served. In the past, moreover, river traffic has been a major factor in keeping the channels open. Oil prospecting is likely to restart once peace has been restored and this may mean that the companies concerned will wish to keep channels clear. However, if discharges drop to the low figures prior to 1961, the canal could become too shallow for commercial traffic and for the movement of fisheries barges.

The canal will in many areas drive a barrier between wet season villages and dry season grazing grounds along the river channels and therefore dislocate the pastoral cycle. Many people living east of the canal will have to cross it with their livestock when regrowth from rain-flooded grasslands is exhausted and they have to move westwards to the river-flooded grasslands of the Nile. Reinforced structures at various points along the canal are needed to facilitate the crossing of livestock without damage to the embankments and to provide suitably designed boats more efficient than the usual 'dug-out' canoe. Crossing the canal will present a massive logistical problem and besides, raises questions of land ownership among those who may need to cross the canal and cross each others' territory in order to do so.

There exists a kind of 'Jonglei Controversy'. The criticism of the environmentalists are many but can be segregated into charges that the Jonglei Canal will drastically affect climate, groundwater recharges, silt and water quality, the destruction of fish and changes in the lifestyle of the Nilotic people. However, other studies claim that the positive effects will counterbalance by far the negative effects. As is the case with the Hadejia-Nguru wetlands understanding and prediction of the impacts is very difficult. However, without such understanding and prediction, effective planning and management is impossible.

Regional aspects of environmental impacts and 'hot spots'

This section summarizes the regional outlook for the main African sub-regions with regard to the impact of irrigation on the environment. For each of the sub-regions, environmentally salient features, particularly in relation to irrigation development issues, are presented. Wherever possible, environmental 'hot spots' are identified and described.

The arid North African sub-region

The North African sub-region lies in arid or semi-arid zones where the water resources are minimal and where evaporation and seepage losses are very large. The sub-region includes two main zones: the Nile Basin in the east and the western part (Morocco, Algeria and Tunisia). This ecological zone is fragile and agricultural production is regulated by alternating periods of water surpluses and deficiencies. Irrigation is the main alternative to cover the food requirements of the increasing population per unit of agricultural land.

The 'hot spots' of the sub-region are the main rivers located in the arid zones threatened by the irrigation-induced salinization of the soils and more generally the degradation of irrigated lands resulting from poor irrigation management and practices. Field drainage and removal of drained water from the irrigated zone is necessary to limit the risk of soil degradation and salinization.

Establishing field drainage is costly, as is the provision of a main drainage network. Moreover, disposal of drainage water represents a major problem. The concentration of salt increases gradually from upstream to downstream as a result of the drainage water inflows.

In the Mediterranean coastal zone, reduction of flow systematically induces sea water intrusion problems.

The Sudano-Sahelian Belt

The natural capital of this ecological zone is the most fragile, evincing most of the negative effects of irrigation projects on the environment due to the poor soils, extremely variable rainfall and high risk of drought.

Soil degradation has substantially increased the risk of desertification because of mutually reinforcing factors including: loss of organic matter and nutrients, soil structure deterioration and surface crusting, which in turn decreases water infiltration and retention, aggravated by the irrigation-induced salinization.

The environmental degradation has been both a cause and a consequence of poverty, with the Sudano-Sahelian belt comprising some of the poorest countries of the world.

Sudano-Sahelian societies face a formidable challenge which makes the whole belt an environmental 'hot spot'. Pressure is likely to be high on the river valleys and major wetlands such as the Niger Inner Delta in Mali, the wetland and flood plains in the Lake Chad basin (Hadejia-Nguru, Yaere) and the Sudd swamps in the Nile basin in southern Sudan.

Humid West Africa

The natural capital of this ecological zone is relatively favourable in terms of climatic conditions: high and regular rainfall, soils of reasonable quality. However, the high population growth during the last decades has placed the environment under serious stress. About half of the total land area is cultivated under reasonably good conditions with a much lower climatic risk than in the Sahel. Forest land has shrunk to less than a third of the total area, and what remains is decreasing at an alarming rate of 1%, the fastest rate in tropical Africa.

The environment of the fragile coastal ecosystems is also threatened by industrial and urban development with increasing pollution levels particularly in the Niger delta of Nigeria. A major part of the biodiversity capital of the sub-region is at risk. Any upstream irrigation project requires special care in order to avoid negative impacts on the wetlands, mangroves and lagoons located in the coastal zone in the Guinea Gulf, from Guinea Bissau to the Niger delta in Nigeria. This zone is likely to become a continuous urban megalopolis with a population of over 50 million people on 500 km of coastal line. The current development of private small-scale irrigation projects, using groundwater for horticultural crops, could contribute to increase the intrusion of saline waters due to overexploitation of coastal aquifers.

The Congo/Zaire Basin

About three-quarters of the total surface area of the sub-region would theoretically be cultivable but a major part of it is under tropical rainforest. Overall pressure to clear the rainforest is still relatively low, except at the periphery of the sub-region where it interfaces with areas of high population density.

Land currently cultivated represents about 15% of the total area. Agricultural activities are relatively less important in the sub-region compared to the rest of the continent. They are focused on supplying a growing urban market and on permanent plantations.

Irrigated areas are marginal compared to the huge potential of land and surface waters. The irrigation development will have a minimal impact on the environment. Global environmental problems faced by the Congo/Zaire basin are less severe than those of the other sub-regions, although its future development will present a serious challenge. In particular, countries need to preserve the primary rainforest for global biodiversity and climatic reasons.

East Africa

The good soils in the eastern African highlands have favoured the development of intensive agriculture, although soils require conservation measures because of steep slopes. Less favourable lands are cultivated under arid and semi-arid conditions. Forests cover less than 20% of the total area of the sub-region. Due to land scarcity, the primary rainforests with their unique biodiversity are at risk.

Due to the pressure of population on arable lands, the environment is at risk particularly in Kenya with a ratio of 0.2 ha per person. To cover the food deficits, areas under irrigation are increasing. Permanent intensive cropping is the current pattern in favourable highlands, but degradation is high under low-input technology and without adequate erosion control measures.

In Tanzania, the central area and the Lake Victoria region represent areas of high population density and areas with a reported high degree of land degradation.

Ethiopia has a very large water resources potential. The development of this resource has been impeded for decades, first by agreements made by colonial powers and then by political instability. The Ethiopian Blue Nile and other tributaries contribute over 80 % of the water in Sudan and Egypt (see Chapter 6, section: *the Nile basin*). The mobilization of this potential would have to take into account environmental and basin issues to mitigate the impact on downstream users.

Southern Africa

The natural capital of the sub-region is very rich in terms of biodiversity and production potential, although large areas are under semi-arid and arid conditions with moderate to high risk of drought.

In some countries, particularly South Africa, past policies have had a negative impact on the environment by encouraging agricultural development through high subsidies on farm inputs and irrigation development without stimulating enough soil and water conservation.

Almost half of the total areas of the sub-region is cultivated with reasonably good soils but climatic conditions are highly variable with a risk of recurrent droughts. To mitigate this risk and to cover the food deficit, areas under irrigation are increasing without significant impact on the environment.

SUMMARY OF ENVIRONMENTAL IMPACT HAZARD RELATED TO IRRIGATION DEVELOPMENT

Table 90 attempts to asses the environmental risk related to irrigation development, resulting from salinization and water-related diseases, as well as possible decrease of forestry, fishery and wildlife. Risk has been classified into three categories: serious, moderate or low, and presented for each basin and each environmental topic. It should be stressed that, although some major trends may be identified, environmental impact is usually of local nature and can vary considerably within a basin.

TABLE 90
Environmental impact assessment of irrigation, by basin

Basin	Irrigation potential (1000 ha)	Environmental impact hazard				
		Salinity	Health	Forest	Fishery	Wildlife
Senegal river	420	+ + +	+ +	+	+	+
Niger river	2 817	+ + +	+ +	+	+ +	+ +
Lake Chad	1 163	+ + +	+ +	+	+ +	+ +
Nile river	8 000	+ + +	+	+	+	+ +
Rift Valley	844	+	+ +	+	+	+
Shebelli-Juba	351	+ + +	+	+	+	+
Congo/Zaire river	9 800	+	+	+ +	+	+
Zambezi river	3 160	+ +	+ +	+	+	+
Okavango	208	+ +	+	+	+	+ + +
Limpopo river	295	+ +	+ +	+	+	+
Orange river	390	+ +	+	+	+	+
South interior	54	+ + +	+	+	+	+
North interior	71	+ + +	+	+	+	+
Mediterranean Coast	850	+ + +	+	+	+	+
North West Coast	1 200	+ + +	+	+	+	+
West Coast	5 113	+	+ +	+	+	+
West Central Coast	835	+	+ +	+ +	+	+
South West Coast	1 808	+ +	+ +	+	+	+ +
South Atlantic Coast	84	+ +	+	+	+	+
Indian Ocean Coast	1 500	+	+	+	+	+
East Central Coast	1 928	+	+ +	+	+	+
North East Coast	78	+ +	+	+	+	+
Madagascar	1 500	+	+ +	+	+	+
Islands	35	+ +	+	+	+	+
Total	42 504					

+ + + : serious
+ + : moderate
+ : low or nil

Chapter 8

General results and conclusions

This report describes the different steps leading to the assessment of the irrigation potential for Africa, as presented in a schematic way in Figure 1. It concentrates mainly on the physical factors controlling irrigation potential - land and water - and only deals with renewable water resources. However, the country studies used in the assessment may implicitly include some assumptions on a reasonable level of investment and allow for other constraints such as environmental and social factors, or the use of non-renewable water resources.

The African continent was divided into 24 major hydrologic units, or major basin groups, as presented in Figure 2 and Table 9. These 24 units were combined with the 53 African countries to obtain 136 basic land units, which were the basis for all calculations and for the information gathered and analysed in this study. These 136 basic units are presented in Tables 1 and 2.

PHYSICAL RESOURCES

Land

The soil and terrain suitability for surface irrigation is presented in the Figures 3 to 5 and Tables 4 and 5.

Figure 3 shows the soil and terrain suitability per type of crop (rice and upland crops, rice being given priority for land suitable in both cases). Table 4 and Figure 4 present the total area of land suitable for surface irrigation as a percentage of the area of the basin for each of the 24 major basin groups. Table 5 and Figure 5 present it as a percentage of the area of the country for each of the 53 countries.

The approach used to compute soil suitability for irrigation has its limitations in the fact that it is based on the information obtained from the 1 : 5 000 000 soil map of the world [1]. In particular, the results have proved sensitive to several selection criteria (see Chapter 3), like terrain slope, and no account is taken of distance and elevation of suitable land in relation to water sources. Nonetheless, the results give a fair idea of the distribution of land for irrigation over the continent. Both figures show that the Sahara Desert in northern Africa has the smallest percentage of land suitable for surface irrigation (< 10%). The suitable land of Egypt is concentrated in the Nile Valley and Delta. The southern African region also has relatively little suitable land. The most suitable areas are located in central Africa (part of the Zambezi basin, the Congo/Zaire and upper Nile basins) and in the Shebelli-Juba basin.

Water

The Sahara Desert is the driest region with an average annual rainfall of less than 40 mm (Table 9 and Figure 7). The West Central Coast is the most humid region with an average annual rainfall of almost 1 800 mm, followed by Madagascar (1 700 mm), Congo/Zaire (1 470 mm) and the West Coast (1 435 mm). For all the other major basin groups, average annual rainfall is less than 1 000 mm.

Figure 6 and Table 6 show the internal renewable water resources for the 53 African countries. The differences between arid and humid regions are clearly demonstrated. Total annual internal renewable water resources are less than 50 km^3 for the whole northern African region and almost 2 000 km^3, or 40 times as much, for the central African region, while the surface areas of both regions are more or less the same.

Irrigation water requirements (IWR)

The assessment of irrigation potential, based on land and water resources, can only be done through the assessment of the irrigation water requirements, which is a function of cropping pattern and climate (rainfall and evapotranspiration). For the purposes of the present study a methodology was developed to assess irrigation water requirements in Africa. Twenty-four irrigation cropping pattern zones were defined, being considered homogeneous in terms of crop calendar, cropping intensity and irrigation efficiency (Table 7 and Figure 8). Irrigation water requirements were computed for different scenarios using the climate data from the FAOCLIM cd-rom and the FAO CROPWAT model in combination with GIS. The results are presented in the Figures 10 and 11. A total of 84 homogeneous irrigation water requirement zones were defined. Table 8 summarizes the data obtained for each of the 84 zones.

Comparing the results with figures available from country studies shows that the methodology manages to assess regional estimates of IWR relatively well. Discrepancies with country studies find their origin mostly in the cropping pattern, cropping intensity and irrigation efficiency scenarios.

REVIEW OF EXISTING INFORMATION ON IRRIGATION POTENTIAL

The review and compilation of existing information on irrigation potential is the main component of this study. Most of the irrigation potential studies are based on physical criteria, but implicitly account for technical and economic considerations by concentrating on areas where irrigation is economically feasible (market, demand) and does not present technical difficulties (access to land and water). All the information was cross-checked with the results of the studies on soil and terrain suitability, water resources and irrigation water requirements and completed where necessary.

Maps 1 to 22 show the information collected on annual discharges for each major basin group. It is important to stress that the review concentrated mainly on surface water resources. In arid regions, where the use of groundwater for irrigation purposes already plays an important role, groundwater was considered in the study. Only renewable groundwater was taken into consideration and not the fossil water resources. This choice can lead to considerable discrepancies for countries which include fossil water in their computation of irrigation potential. Libya, for instance, estimates its irrigation potential at 750 000 ha, while this study mentions only 40 000 ha, based on renewable water resources. It was beyond the scope of this study to discuss the use of fossil water and it was removed for the sake of homogeneity in computation.

In most country studies the need for sharing water between agriculture, industries, communities and other uses is taken into account in the assessment of irrigation potential. The fact remains that the part allocated to agriculture depends on assumptions on the rate of development of the other sectors.

Tables 88 and 89 show the degree of availability of information on water resources and irrigation potential by country and by basic unit. Reasonably detailed information was available for about 30% of the countries. In general, little information was available for humid countries. This may be linked to the fact that countries with limited water resources need to plan more carefully the use of these resources and their distribution over the different sectors than the more humid countries. This results in a larger availability of Water Master Plans and irrigation development in the drier countries.

Table 91 shows the irrigation potential estimates resulting from the country studies, amounting to a total of 46.7 million hectares. The study by basin, using the same data as at country level, gives a total of 42.5 million hectares. The difference of 4.2 million hectares is explained by three main factors:

- different countries within a river basin considered the same water as being available for their use (double counting);
- several arid countries included the use of fossil water in their estimates (e.g. Algeria, Libya, Tunisia) while this study considers only renewable water resources;
- several countries considered lower irrigation water requirements in the computation of irrigation potential than recommended in this study (e.g. Algeria, Tunisia).

CONCLUSIONS

Figure 14 shows the basin groups where water is abundant and those with a risk of water scarcity, if the whole potential is developed. Once again, it should be stressed that in the figures each major basin group was considered as one entity, while in reality water may be scarce only in part of the basin. For example, in the Niger basin, few problems will exist in the Benue tributary basin, but problems may arise in the main Niger River region straddling Mali and Niger. For the Indian Ocean Coast, problems may arise in the basins shared between Swaziland, South Africa and Mozambique, but not in the basins that are located entirely within South Africa. On the other hand, for the Zambezi basin as a whole no water problems exist according to the figures, but in the upper part, occupied by the Chobe tributary basin which is shared between Angola, Namibia, Zambia and Botswana, water problems may arise. Details are given in Chapter 6.

Figure 15 presents the distribution of the 42.5 million hectares of irrigation potential over the different basins. Figures for the basins in arid areas have to be considered as upper limits from the point of view of water availability, while for some humid basins, where neither water nor land is a limiting factor, other factors (mostly economic) have been taken into account.

TABLE 91
Irrigation potential by country and by basin in hectares

Basin / Country	Senegal	Niger	Lake Chad	Nile	Rift	Sheb-Juba	Congo/Zaire	Zambezi	Okavango	Limpopo	Orange	South Int.	North Int.	Med.Coast	N.W.Coast	W.Coast	W.C.Coast	S.W.Coast	S.A.Coast	I.O.Coast	E.C.Coast	N.E.Coast	Madagascar	Islands	Total for country	Basin / Country
Country	(1)	(2)	(3)	(4)	(5)	(6)	(7)	(8)	(9)	(10)	(11)	(12)	(13)	(14)	(15)	(16)	(17)	(18)	(19)	(20)	(21)	(22)	(23)	(24)		
Algeria		0	0										125 200	385 100	0										510 300	Algeria
Angola							900 000	700 000	200 000			50 000					50 000	1 800 000							3 700 000	Angola
Benin		100 000														200 000									300 000	Benin
Botswana								1 080	6 060	5 000	0	2 500													14 640	Botswana
Burkina Faso		5 000														159 460									164 460	Burkina Faso
Burundi				80 000			105 000																		185 000	Burundi
Cameroon		20 000	100 000				50 000										120 000								290 000	Cameroon
Cape Verde																								2 990	2 990	Cape Verde
Central African Rep.			500 000				1 400 000																		1 900 000	Central African Rep.
Chad		0	835 000										0												835 000	Chad
Comoros																								300	300	Comoros
Congo							255 000										85 000								340 000	Congo
Cote d'Ivoire		50 000														425 000									475 000	Cote d'Ivoire
Djibouti					450																	550			1 000	Djibouti
Egypt				4 420 000									0	0								0			4 420 000	Egypt
Equatorial Guinea																	30 000								30 000	Equatorial Guinea
Eritrea				150 000	0																	37 500			187 500	Eritrea
Ethiopia				2 220 000	790 000	627 300																0			3 637 300	Ethiopia
Gabon																	440 000								440 000	Gabon
Gambia																80 000									80 000	Gambia
Ghana																1 900 000									1 900 000	Ghana
Guinea	5 000	185 000														330 000									520 000	Guinea
Guinea Bissau																281 290									281 290	Guinea Bissau
Kenya				180 000	52 500	9 460															111 100				353 060	Kenya
Lesotho											12 500														12 500	Lesotho
Liberia																600 000									600 000	Liberia
Libya													0	40 000											40 000	Libya
Madagascar																							1 500 000		1 500 000	Madagascar
Malawi								160 900													1 000				161 900	Malawi
Mali	10 000	556 000											0			0									566 000	Mali
Mauritania	165 000												0		0										165 000	Mauritania
Mauritius																								20 000	20 000	Mauritius
Morocco+W.Sahara													60 000	380 000	1 200 000										1 640 000	Morocco+W.Sahara
Mozambique								1 700 000		148 000										368 000	856 000				3 072 000	Mozambique
Namibia								11 000	2 000		25 000	400						7 900	1 000						47 300	Namibia
Niger		222 000	48 000										0												270 000	Niger
Nigeria		1 678 510	502 000													50 000	100 000								2 330 510	Nigeria
Rwanda				150 000			9 000																		159 000	Rwanda
Sao Tome & Principe																								10 700	10 700	Sao Tome & Principe
Senegal	240 000															100 000									340 000	Senegal
Seychelles																								1 000	1 000	Seychelles
Sierra Leone																807 000									807 000	Sierra Leone
Somalia						230 000															0	10 000			240 000	Somalia
South Africa										131 500	352 500								83 200	877 800					1 445 000	South Africa
Sudan			4 000	2 750 000	0								0									30 000			2 784 000	Sudan
Swaziland																				93 220					93 220	Swaziland
Tanzania				30 000	1 060		0	0													959 360				990 420	Tanzania
Togo																180 000									180 000	Togo
Tunisia													40 000	523 000											563 000	Tunisia
Uganda				202 000	0																				202 000	Uganda
Zaire				10 000			6 980 000										10 000								7 000 000	Zaire
Zambia							101 000	422 000																	523 000	Zambia
Zimbabwe								165 400		10 900		1 100								211 000					388 400	Zimbabwe
Sum of countries (C)	420 000	2 816 510	1 989 000	10 192 000	844 010	866 760	9 800 000	3 160 380	208 060	295 400	390 000	54 000	225 200	1 328 100	1 200 000	5 112 750	835 000	1 807 900	84 200	1 550 020	1 927 460	78 050	1 500 000	34 990	46 719 790	Sum of countries (C)
Total for basin (B)	420 000	2 816 510	1 163 200	8 000 000	844 010	351 460	9 800 000	3 160 380	208 060	295 400	390 000	54 000	71 000	850 000	1 200 000	5 112 750	835 000	1 807 900	84 200	1 500 000	1 927 460	78 050	1 500 000	34 990	42 504 370	Total for basin (B)
Difference: (C) - (B)	0	0	825 800	2 192 000	0	515 300	0	0	0	0	0	0	154 200	478 100	0	0	0	0	0	50 020	0	0	0	0	4 215 420	Difference: (C) - (B)
Basin as % countries	100%	100%	58%	78%	100%	41%	100%	100%	100%	100%	100%	100%	32%	64%	100%	100%	100%	100%	100%	97%	100%	100%	100%	100%	91%	Basin as % countries
Basin	Senegal	Niger	Lake Chad	Nile	Rift	Sheb-Juba	Congo/Zaire	Zambezi	Okavango	Limpopo	Orange	South Int.	North Int.	Med.Coast	N.W.Coast	W.Coast	W.C.Coast	S.W.Coast	S.A.Coast	I.O.Coast	E.C.Coast	N.E.Coast	Madagascar	Islands	TOTAL	Basin

In Figure 16 irrigation potential is expressed in hectares per km^2 of basin area (or in percentage of basin area). The figure clearly shows that irrigation potential is less in the arid northern, north-eastern and southern regions than in the rest of the continent.

Table 92 compares the soil and terrain suitability figures at basin level with the irrigation potential figures obtained from the literature review. For the continent as a whole, 42.5 million hectares correspond to only 7% of the area with soil and terrain suitable for irrigation (almost 600 million hectares). The basins with the highest percentages are the humid ones, where water is not a limiting factor, and those located in areas with high population densities (Figure 17) and where population pressure leads to a need for maximum agricultural development. Analysing the figures at basic unit level, the part of Egypt lying in the Nile basin is the only area in the whole continent where land is the real limiting factor and where marginal soils have been included in the irrigation potential figure.

TABLE 92
Comparison of the figures on soil and land suitability for irrigation with the figures on irrigation potential, by basin

Basin No.	Basin	Total area of basin (ha)	Land suitable for surface irrigation (ha)	Irrigation potential of basin (ha)	Irrigation potential as % of suitable land
	(1)	*(2)*	*(3)*	*(4)*	*(5) = 100*(4)/(3)*
01	Senegal	48 318 100	3 645 800	420 000	11.52
02	Niger	227 394 600	28 943 400	2 816 510	9.73
03	Lake Chad	238 163 500	36 524 600	1 163 200	3.18
04	Nile	311 236 900	92 019 000	8 000 000	8.69
05	Rift Valley	63 759 300	13 946 700	844 010	6.05
06	Shebelli-Juba	81 042 700	25 847 900	351 460	1.36
07	Congo/Zaire	378 905 300	109 815 500	9 800 000	8.92
08	Zambezi	135 136 500	37 632 500	3 160 380	8.40
09	Okavango	32 319 200	6 612 100	208 060	3.15
10	Limpopo	40 186 400	9 736 100	295 400	3.03
11	Orange	89 636 800	14 140 500	390 000	2.76
12	South Interior	64 582 600	15 421 800	54 000	0.35
13	North Interior	580 446 300	48 325 700	71 000	0.15
14	Mediterranean Coast	67 952 500	11 897 700	850 000	7.14
15	North West Coast	67 062 100	12 565 200	1 200 000	9.55
16	West Coast	143 019 600	29 567 400	5 112 750	17.29
17	West Central Coast	70 477 400	16 335 000	835 000	5.11
18	South West Coast	51 620 000	13 792 500	1 807 900	13.11
19	South Atlantic Coast	36 548 500	4 041 900	84 200	2.08
20	Indian Ocean Coast	66 378 500	14 853 200	1 500 000	10.10
21	East Central Coast	102 625 200	24 913 500	1 927 460	7.74
22	North East Coast	72 570 200	11 779 100	78 050	0.66
24	Madagascar	58 704 000	14 497 400	1 500 000	10.35
25	Islands	934 600	105 500	34 990	33.17
	Total	3 029 020 800	596 960 000	42 504 370	7.12

While this study was based on average annual discharges, important seasonal variations do exist. In all cases full development of the whole irrigation potential would require important storage works and collaboration between countries on the management of shared waters within individual basins.

TABLE 93
Irrigation potential, irrigated areas and possibilities for irrigation expansion, by basin

Basin No.	Basin	Total area of basin (ha)	Irrigation potential of basin (ha)	Irrigation potential as % of basin area	Area under irrigation (ha)	Irrigation as % of irrigation potential	Possibility for irrigation expansion (ha)
	(1)	(2)	(3)	(4) = 100*(3)/(2)	(5)	(6) = 100*(5)/(3)	(7) = (3)-(5)
01	Senegal	48 318 100	420 000	0.87	118 150	28.1	301 850
02	Niger	227 394 600	2 816 510	1.24	228 240	8.1	2 588 270
03	Lake Chad	238 163 500	1 163 200	0.49	113 296	9.7	1 049 904
04	Nile	311 236 900	8 000 000	2.57	5 078 604	63.5	2 921 396
05	Rift Valley	63 759 300	844 010	1.32	193 496	22.9	650 514
06	Shebelli-Juba	81 042 700	351 460	0.43	199 000	56.6	152 460
07	Congo/Zaire	378 905 300	9 800 000	2.59	35 767	0.4	9 764 233
08	Zambezi	135 136 500	3 160 380	2.34	146 869	4.6	3 013 511
09	Okavango	32 319 200	208 060	0.64	0	0.0	208 060
10	Limpopo	40 186 400	295 400	0.74	241 381	81.7	54 019
11	Orange	89 636 800	390 000	0.44	302 722	77.6	87 278
12	South Interior	64 582 600	54 000	0.08	250	0.5	53 750
13	North Interior	580 446 300	71 000	0.01	232 500	327.5	- 161 500
14	Mediterran. Coast	67 952 500	850 000	1.25	1 606 700	189.0	- 756 700
15	North West Coast	67 062 100	1 200 000	1.79	1 000 750	83.4	199 250
16	West Coast	143 019 600	5 112 750	3.57	310 883	6.1	4 801 867
17	West Central Coast	70 477 400	835 000	1.18	28 450	3.4	806 550
18	South West Coast	51 620 000	1 807 900	3.50	70 000	3.9	1 737 900
19	South Atlant. Coast	36 548 500	84 200	0.23	84 000	99.8	200
20	Indian Ocean Coast	66 378 500	1 500 000	2.26	862 110	57.5	637 890
21	East Central Coast	102 625 200	1 927 460	1.88	178 610	9.3	1 748 850
22	North East Coast	72 570 200	78 050	0.11	24 574	31.5	53 476
24	Madagascar W+E	58 704 000	1 500 000	2.56	1 087 000	72.5	413 000
25	Islands	934 600	34 990	3.74	30 109	86.1	4 881
	Total	3 029 020 800	42 504 370	1.40	12 173 461	28.6	30 330 909

(6)+(7): In estimating the irrigation potential only renewable water resources are taken into consideration.
If >100% (6) or negative (7) either non-renewable water resources are already being used for irrigation or the quantities of water used per hectare are less than the quantities recommended for the potential area.

The present study concentrated on long-term averages. Climate fluctuations, however, may greatly influence the possibilities for irrigation development. In the Niger basin, for example the 1980s were much drier than earlier years: the average annual discharges of the 1980s were 20 to 40% less than those before the 1980s. The average annual discharge of the White Nile entering Sudan from Uganda during the period 1961-1980 (50 km^3/year) was nearly twice the average annual discharge during the period 1905-1960 (27 km^3/year). The recent drought years in Southern Africa will also lead to different averages depending on the period of reference considered.

Figure 18 and Tables 93 to 95 show the part of irrigation potential by basin which is already under irrigation at present[1]. As expected, it is much lower in the humid regions than in the dry regions. In a few basins, figures higher than 100% mean that the area already under irrigation is larger than the potential. This is due to the fact that either fossil water is used for irrigation (northern Africa) or that the quantity of water used per hectare is less than the

[1] This is the area occupied by irrigation schemes with full or partial control (11.5 million ha), spate irrigation (0.5 million ha) and wetlands and inland valley bottoms that are equipped for water control (0.2 million ha).

TABLE 94
Area under irrigation by country and by basin in hectares

Country / Basin	Senegal (1)	Niger (2)	Lake Chad (3)	Nile (4)	Rift (5)	Sheb-Juba (6)	Congo/Zaire (7)	Zambezi (8)	Okavango (9)	Limpopo (10)	Orange (11)	South Int. (12)	North Int. (13)	Med.Coast (14)	N.W.Coast (15)	W.Coast (16)	W.C.Coast (17)	S.W.Coast (18)	S.A.Coast (19)	I.O.Coast (20)	E.C.Coast (21)	N.E.Coast (22)	Madagascar (23)	Islands (24)	Total for country	Basin / Country
Algeria		0	0										45 000	510 500	0										555 500	Algeria
Angola							2 000	2 000	0			0					1 000	70 000							75 000	Angola
Benin		1 090														9 146									10 236	Benin
Botswana								0	0	1 381	0	0													1 381	Botswana
Burkina Faso		850														23 480									24 330	Burkina Faso
Burundi				0			14 400																		14 400	Burundi
Cameroon		2 500	13 820				1 650										3 000								20 970	Cameroon
Cape Verde																								2 779	2 779	Cape Verde
Central African Rep.			135				0																		135	Central African Rep.
Chad		0	14 020										0												14 020	Chad
Comoros																								130	130	Comoros
Congo							217										0								217	Congo
Cote d'Ivoire		0														72 750									72 750	Cote d'Ivoire
Djibouti					100																	574			674	Djibouti
Egypt				3 078 000									0	168 000								0			3 246 000	Egypt
Equatorial Guinea																	-								-	Equatorial Guinea
Eritrea				15 124	0																	13 000			28 124	Eritrea
Ethiopia				23 160	166 396	0																0			189 556	Ethiopia
Gabon																	4 450								4 450	Gabon
Gambia																1 670									1 670	Gambia
Ghana																6 374									6 374	Ghana
Guinea	0	1 000														91 880									92 880	Guinea
Guinea Bissau																17 115									17 115	Guinea Bissau
Kenya				6 000	27 000	0															33 610				66 610	Kenya
Lesotho											2 722														2 722	Lesotho
Liberia																2 100									2 100	Liberia
Libya													150 000	320 000											470 000	Libya
Madagascar																							1 087 000		1 087 000	Madagascar
Malawi								28 000													0				28 000	Malawi
Mali	300	78 320											0			0									78 620	Mali
Mauritania	46 450												2 000		750										49 200	Mauritania
Mauritius																								17 500	17 500	Mauritius
Morocco+W.Sahara													10 000	248 200	1 000 000										1 258 200	Morocco+W.Sahara
Mozambique								20 000		40 000										41 710	5 000				106 710	Mozambique
Namibia								6 142	0		0	0						0	0						6 142	Namibia
Niger		64 480	2 000										0												66 480	Niger
Nigeria		80 000	82 821													50 000	20 000								232 821	Nigeria
Rwanda				2 000			2 000																		4 000	Rwanda
Sao Tome & Principe																								9 700	9 700	Sao Tome & Principe
Senegal	71 400															0									71 400	Senegal
Seychelles																										Seychelles
Sierra Leone																29 360									29 360	Sierra Leone
Somalia						199 000															0	1 000			200 000	Somalia
South Africa										198 000	300 000								84 000	688 000					1 270 000	South Africa
Sudan			500	1 935 200	0								500									10 000			1 946 200	Sudan
Swaziland																				67 400					67 400	Swaziland
Tanzania				10 000	0		0	0													140 000				150 000	Tanzania
Togo																7 008									7 008	Togo
Tunisia													25 000	360 000											385 000	Tunisia
Uganda				9 120	0																				9 120	Uganda
Zaire				0			10 500										0								10 500	Zaire
Zambia							5 000	41 400																	46 400	Zambia
Zimbabwe								49 327		2 000		250								65 000					116 577	Zimbabwe
Sum of countries (C)	118 150	228 240	113 296	5 078 604	193 496	199 000	35 767	146 869	0	241 381	302 722	250	232 500	1 606 700	1 000 750	310 883	28 450	70 000	84 000	862 110	178 610	24 574	1 087 000	30 109	12 173 461	Sum of countries (C)
Potential for basin (B)	420 000	2 816 510	1 163 200	8 000 000	844 010	351 460	9 800 000	3 160 380	208 060	295 400	390 000	54 000	71 000	850 000	1 200 000	5 112 750	835 000	1 807 900	84 200	1 500 000	1 927 460	78 050	1 500 000	34 990	42 504 370	Potential for basin (B)
(C) as % of (B)	28.1%	8.1%	9.7%	63.5%	22.9%	56.6%	0.4%	4.6%	0.0%	81.7%	77.6%	0.5%	327.5%	189.0%	83.4%	6.1%	3.4%	3.9%	99.8%	57.5%	9.3%	31.5%	72.5%	86.1%	28.6%	(C) as % of (B)
Basin	Senegal	Niger	Lake Chad	Nile	Rift	Sheb-Juba	Congo/Zaire	Zambezi	Okavango	Limpopo	Orange	South Int.	North Int.	Med.Coast	N.W.Coast	W.Coast	W.C.Coast	S.W.Coast	S.A.Coast	I.O.Coast	E.C.Coast	N.E.Coast	Madagascar	Islands	TOTAL	Basin

TABLE 95

Area under irrigation as percentage of irrigation potential by country and by basin

Basin / Country	Senegal (1)	Niger (2)	Lake Chad (3)	Nile (4)	Rift (5)	Sheb-Juba (6)	Congo/Zaire (7)	Zambezi (8)	Okavango (9)	Limpopo (10)	Orange (11)	South Int. (12)	North Int. (13)	Med.Coast (14)	N.W.Coast (15)	W.Coast (16)	W.C.Coast (17)	S.W.Coast (18)	S.A.Coast (19)	I.O.Coast (20)	E.C.Coast (21)	N.E.Coast (22)	Madagascar (23)	Islands (24)	Country (C)	Basin / Country
Algeria		0.0	0.0										35.9	139.8	0.0										108.9	Algeria
Angola							0.2	0.3	0.0			0.0					2.0	3.9							2.0	Angola
Benin		1.1														4.6									3.4	Benin
Botswana								0.0	0.0	27.6	0.0	0.0													9.4	Botswana
Burkina Faso		17.0														14.7									14.8	Burkina Faso
Burundi				0.0			13.7																		7.8	Burundi
Cameroon		12.5	13.8				3.3										2.5								7.2	Cameroon
Cape Verde																								92.9	92.9	Cape Verde
Centr. Afr. Rep.			0.0				0.0																		0.0	Centr. Afr. Rep.
Chad		0.0	1.7										0.0												1.7	Chad
Comoros																								43.3	43.3	Comoros
Congo							0.1										0.0								0.1	Congo
Cote d'Ivoire		0.0														17.1									15.3	Cote d'Ivoire
Djibouti					22.2																	104.4			57.4	Djibouti
Egypt				69.6									0.0	> 100.0								0.0			73.4	Egypt
Equat. Guinea																									0.0	Equat. Guinea
Eritrea				10.1	0.0																	34.7			13.9	Eritrea
Ethiopia				1.0	21.1	0.0																0.0			5.2	Ethiopia
Gabon																	1.0								1.0	Gabon
Gambia																2.1									2.1	Gambia
Ghana																0.3									0.3	Ghana
Guinea	0.0	0.5														27.8									17.9	Guinea
Guinea Bissau																6.1									6.1	Guinea Bissau
Kenya				3.3	51.4	0.0															30.3				18.9	Kenya
Lesotho											21.8														21.8	Lesotho
Liberia																0.4									0.4	Liberia
Libya													> 100.0	800.0											1175.0	Libya
Madagascar																							72.5		72.5	Madagascar
Malawi								17.4													0.0				17.3	Malawi
Mali	3.0	14.1											0.0			0.0									13.9	Mali
Mauritania	28.2												> 100.0		> 100.0										29.8	Mauritania
Mauritius																								87.5	87.5	Mauritius
Mor.+W.Sah.													16.7	65.3	83.3										76.7	Mor.+W.Sah.
Mozambique								1.2		27.0										11.3	0.6				3.5	Mozambique
Namibia								55.8	0.0		0.0	0.0						0.0	0.0						13.0	Namibia
Niger		29.0	4.2										0.0												24.6	Niger
Nigeria		4.8	16.5													100.0	20.0								10.0	Nigeria
Rwanda				1.3			22.2																		2.5	Rwanda
Sao Tome & Pr.																								90.7	90.7	Sao Tome & Pr.
Senegal	29.8															0.0									21.0	Senegal
Seychelles																									0.0	Seychelles
Sierra Leone																3.6									3.6	Sierra Leone
Somalia						86.5															0.0	10.0			83.3	Somalia
South Africa										150.6	85.1								101.0	78.4					87.9	South Africa
Sudan			12.5	70.4	0.0								> 100.0									33.3			69.9	Sudan
Swaziland																				72.3					72.3	Swaziland
Tanzania				33.3	0.0		0.0	0.0													14.6				15.1	Tanzania
Togo																3.9									3.9	Togo
Tunisia													62.5	68.8											68.4	Tunisia
Uganda				4.5	0.0																				4.5	Uganda
Zaire				0.0			0.2										0.0								0.2	Zaire
Zambia							5.0	9.8																	8.9	Zambia
Zimbabwe								29.8		18.3		22.7								30.8					30.0	Zimbabwe
Basin (B)	28.1	8.1	9.7	63.5	22.9	56.6	0.4	4.6	0.0	81.7	77.6	0.5	327.5	189.0	83.4	6.1	3.4	3.9	99.8	57.5	9.3	31.5	72.5	86.1	28.6	Basin (B)
Basin	Senegal	Niger	Lake Chad	Nile	Rift	Sheb-Juba	Congo/Zaire	Zambezi	Okavango	Limpopo	Orange	South Int.	North Int.	Med.Coast	N.W.Coast	W.Coast	W.C.Coast	S.W.Coast	S.A.Coast	I.O.Coast	E.C.Coast	N.E.Coast	Madagascar	Islands	TOTAL	Basin

C (country level) = area under irrigation as % of irrigation potential of the country; B (basin level) = area under irrigation as % of irrigation potential of basin irrigation water requirement figure used in the computation of the irrigation potential (both northern and southern Africa). Figure 19 shows the remaining possibilities for irrigation expansion, which are greatest in the humid regions and smallest or even negative (for the reasons explained above) in the dry regions.

Table 96 compares the potential and current situation of areas under irrigation, irrigation water requirements and total water use for each major basin group. The current water use in column 7, available from AQUASTAT [21a], is based on information provided by the countries. It must be noted here, that methods for calculating agricultural water withdrawal vary from country to country, which may lead to some inconsistencies. For example, Somalia probably only included the water withdrawal for the full and partial control irrigation schemes, estimated at 50 000 ha and not the water withdrawal for the 150 000 ha of spate irrigation. This may explain the low figure of current irrigation water requirement (4 000 m^3/ha per year) for the Shebelli-Juba basin, calculated by dividing the total current water use (column 5) by the total area under irrigation (column 4). Also, the area under irrigation in column 4 represents the area *equipped for irrigation*. It may be possible, that only part of the equipped area in the basins is actually irrigated, which again may lead to lower figures for irrigation water requirement, if dividing total water use by the *equipped* area. However, it is estimated that most figures in this table reflect reasonably well the situation at basin and continental level. The potential water requirement for the whole continent is estimated at about 614 km^3 per year, which is almost five times the current agricultural water use, estimated at 128 km^3 per year.

TABLE 96
Comparison of potential and current situation on areas under irrigation and water requirements by basin

Basin	Potential situation			Current situation			Actual water use as % of potential water req. *
	total area (ha)	total water requirement (km^3/year)	gross water req. per ha (m^3/ha.yr)	total area (ha)	total water use (km^3/year)	gross water use per ha (m^3/ha.yr)	
	(1)	*(2)*	*(3) = 10^9*(2)/(1)*	*(4)*	*(5)*	*(6) = 10^9*(5)/(4)*	*(7) = 100*(5)/(2)*
Senegal river	420 000	14.37	34 000	118 150	2.67	22 500	18.6
Niger river	2 816 510	55.02	19 500	228 240	2.43	10 500	4.4
Lake Chad	1 163 200	16.53	14 000	113 296	1.00	9 000	6.0
Nile river	8 000 000	100.00	12 500	5 078 604	60.78	12 000	60.8
Rift Valley	844 010	7.91	9 500	193 496	2.08	10 500	26.3
Shebelli-Juba	351 460	5.00	14 000	199 000	0.78	4 000	15.6
Congo/Zaire river	9 800 000	158.86	16 000	35 767	0.68	19 000	0.4
Zambezi river	3 160 380	37.30	12 000	146 869	2.67	18 000	7.2
Okavango	208 060	1.45	7 000	0	0.00	0	0.0
Limpopo river	295 400	3.45	11 500	241 381	1.77	7 500	51.3
Orange river	390 000	4.88	12 500	302 722	2.29	7 500	46.9
South Interior	54 000	0.28	5 000	250	0.00	8 000	0.7
North Interior	71 000	1.00	14 000	232 500	1.82	8 000	182.0
Mediterranean Coast	850 000	8.19	9 500	1 606 700	12.11	7 500	147.9
North West Coast	1 200 000	12.00	10 000	1 000 750	8.11	8 000	67.6
West Coast	5 112 750	90.09	17 500	310 883	2.52	8 000	2.8
West Central Coast	835 000	11.00	13 000	28 450	0.20	7 000	1.8
South West Coast	1 807 900	24.43	13 500	70 000	0.34	5 000	1.4
South Atlantic Coast	84 200	0.85	10 000	84 000	0.63	7 500	74.1
Indian Ocean Coast	1 500 000	15.99	10 500	862 110	6.57	7 500	41.1
East Central Coast	1 927 460	20.94	11 000	178 610	1.79	10 000	8.5
North East Coast	78 050	1.11	14 000	24 574	0.21	8 500	18.9
Madagascar	1 500 000	23.25	15 500	1 087 000	16.14	15 000	69.4
Islands	34 990	0.32	9 000	30 109	0.30	10 000	93.8
Total	42 504 370	614.22	14 500	12 173 461	127.89	10 500	20.8

*: A figure of > 100% means, that at present fossil water is used for irrigation, while the irrigation potential figure is based on the use of renewable water resources only.

The area under irrigation (12.2 million ha) is less than 30% of the irrigation potential (42.5 million ha). Although considerable potential for future expansion still exists, several observations have to be made:

- Over 60% of the irrigation potential is located in humid regions, and almost 25% in the Congo/Zaire basin alone. These are the regions where potential for rainfed agriculture is also high and where irrigation is mainly supplementary irrigation. It is also in these regions that irrigation is at present least developed. Out of a potential of 9.8 million hectares for the Congo/Zaire basin, only 1% has been developed so far.

- In the regions where irrigation is most important for agriculture, between 60% and more than 100% of the potential (when considering only renewable water resources) is already irrigated. Most of the areas presenting the best potential are already under irrigation. New development will typically require higher investments in terms of water regulation or transportation, or will take place on less productive soils.

- Out of the 12.2 million hectares considered under irrigation, it is estimated that over 50% need rehabilitation if they are to be managed to the maxiumum of their potential. As some irrigation systems function badly or not at all, rehabilitation could also contribute to improving irrigation performance. However, innovative thinking and research are needed to avoid the same failures recurring in the future. Farmers should be actively involved, as they have valuable knowledge regarding deficiencies of the existing system. Investments in rehabilitation and modernization should be used to provide incentives for management reform in existing bureaucratically-run irrigation projects.

The figures of this study concern mainly the physical potential with some considerations about technical and economic feasibility. It is impossible to integrate complex issues, including economic, political, social and environmental aspects, into a purely quantitative assessment exercise at the scale of a continent. However, in addition to the physical resources, socio-economic and environmental as well as political considerations will determine the real possibilities for irrigation development and the choices to be made on the use of water in a river basin, as presented in a schematic way in Figure 1.

Any economic activity, and changes to it, can have different impacts upon men, women and children. This also applies to the water development sector and, thus, any assessment of the net development contribution to a social system such as an irrigation system requires a social analysis, where gender analysis is one component for ensuring that policy and projects are effective, efficient and will have a significant development impact [8].

FIGURE 14
Irrigation potential in Africa: water as a limiting factor

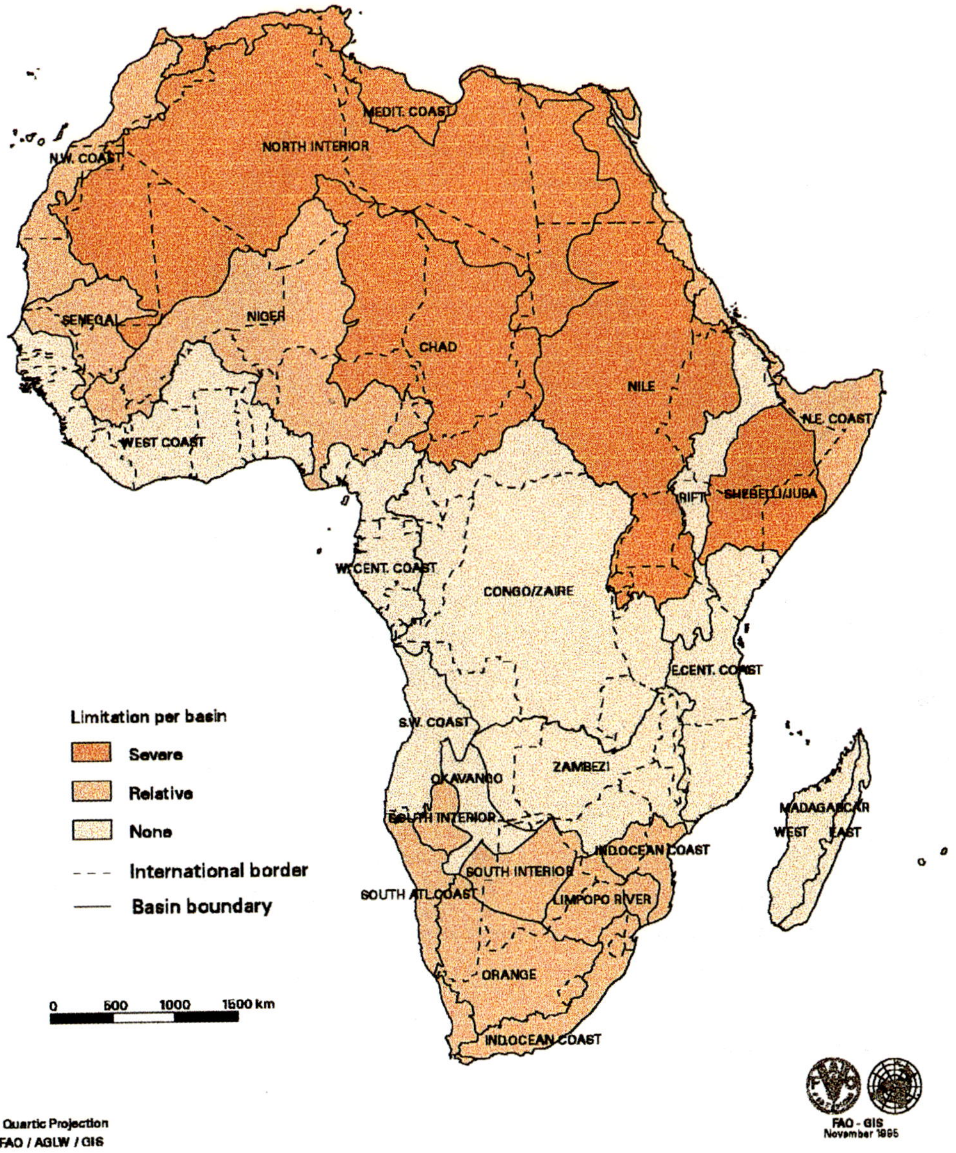

Flat Polar Quartic Projection
Sources: FAO / AGLW / GIS

FIGURE 15
Irrigation potential in Africa (in millions of hectares per basin)

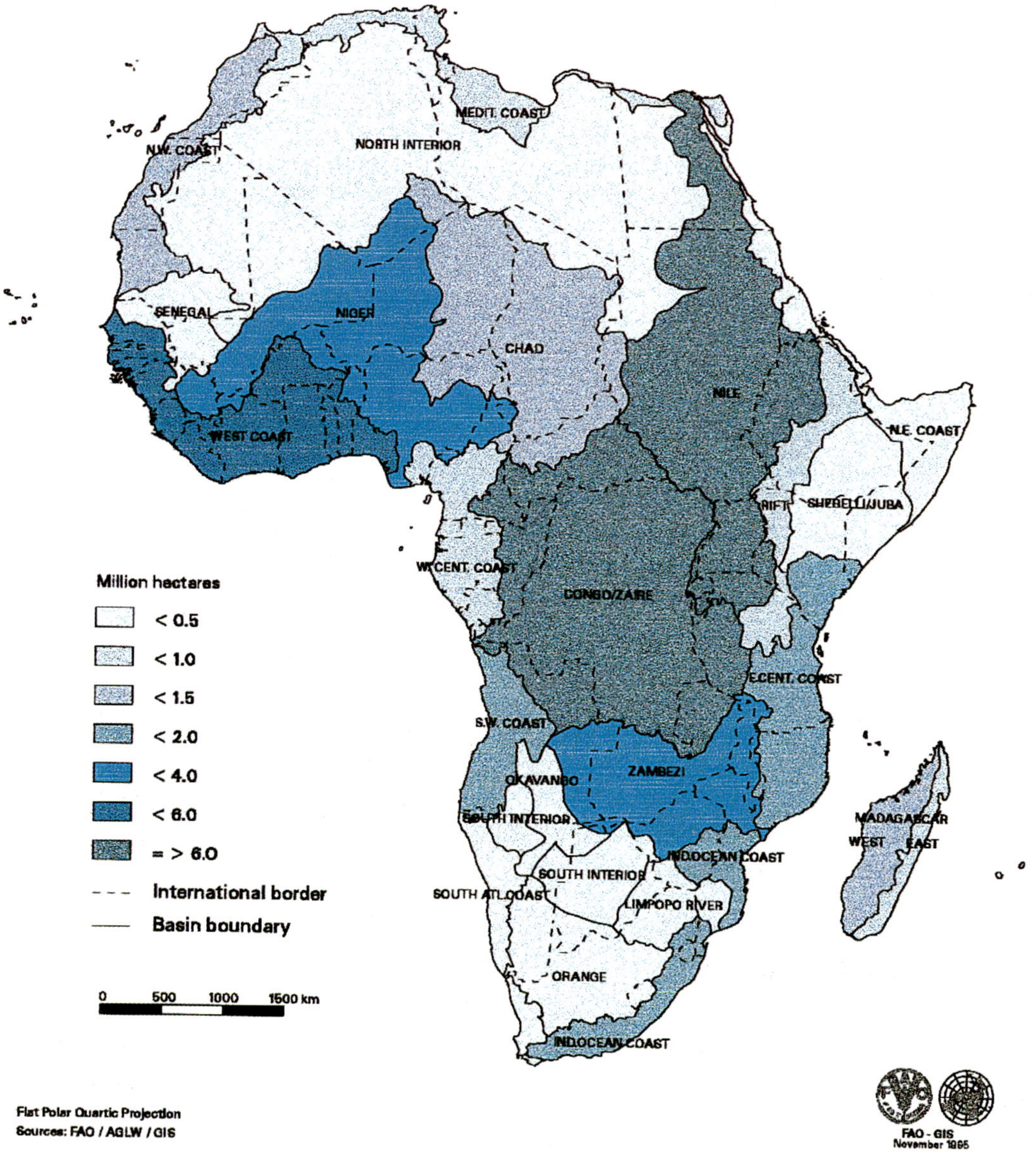

FIGURE 16
Irrigation potential in Africa as % of basin area

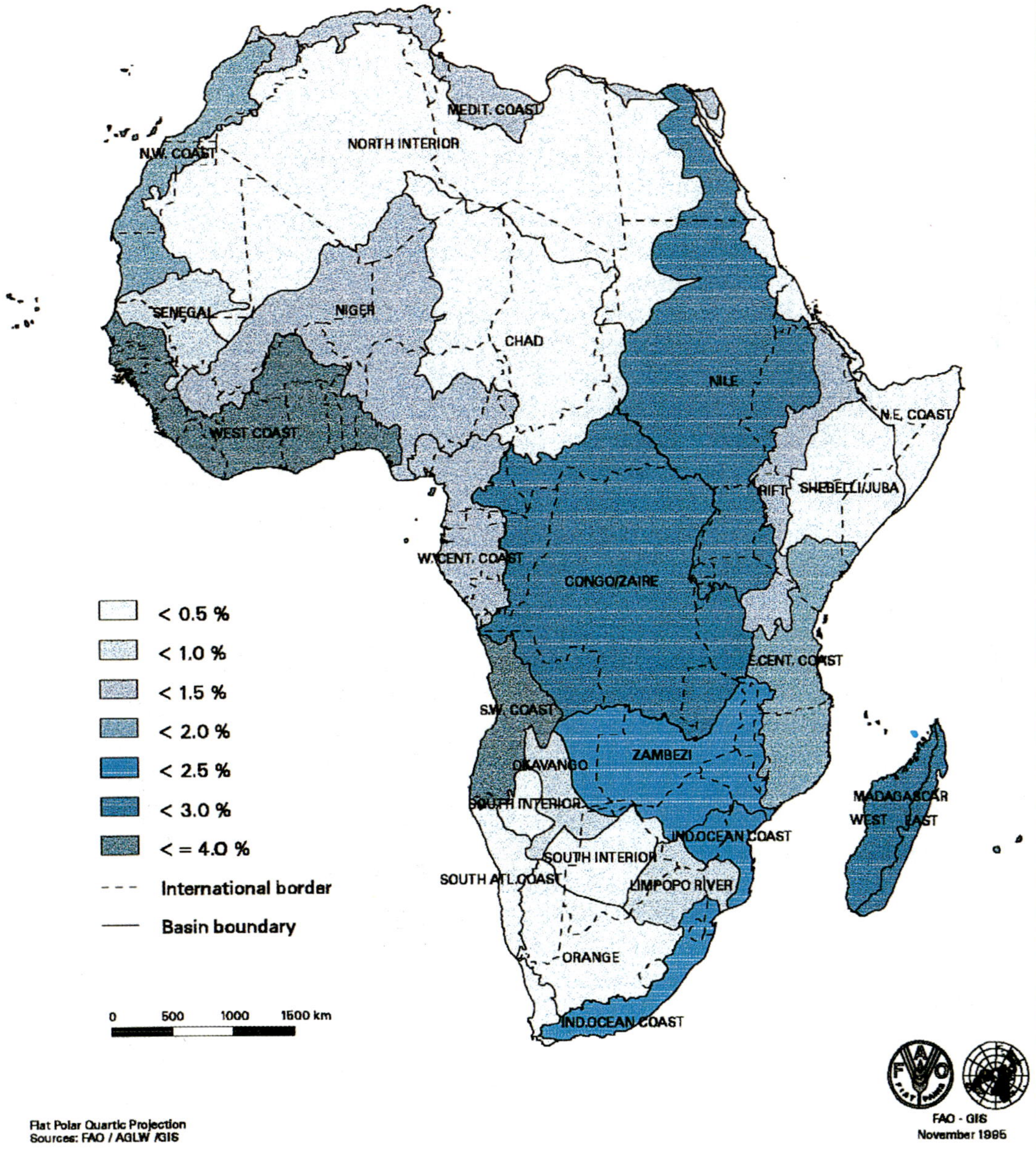

FIGURE 17
Africa: density of population by administrative unit (estimated for 1994)

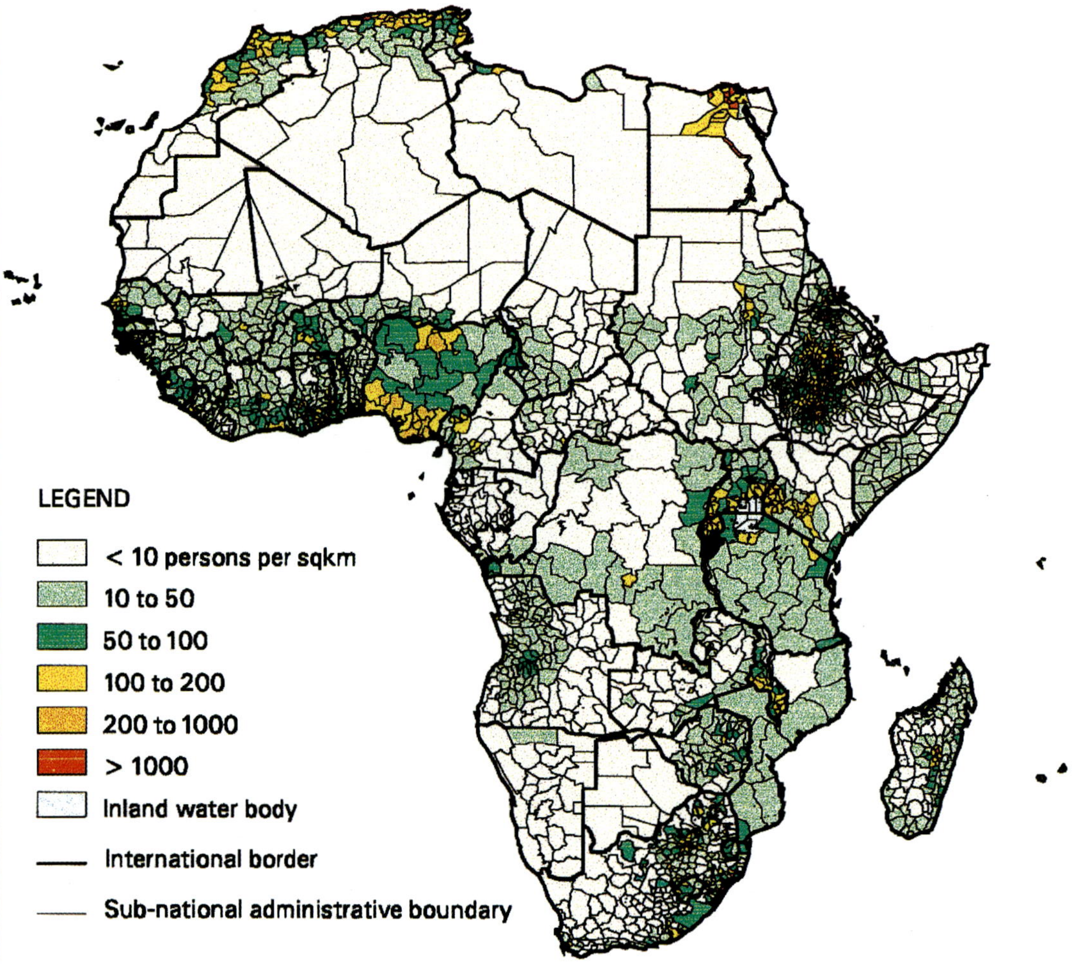

Scale 1 : 73 200 000 approx. at the equator
GEOGRAPHIC PROJECTION (Lat/Long)

Source: AGLS, FAO and University of California, Santa Barbara
September 1994

FAO - GIS

FIGURE 18
Irrigation in Africa (as percentage of basin potential)

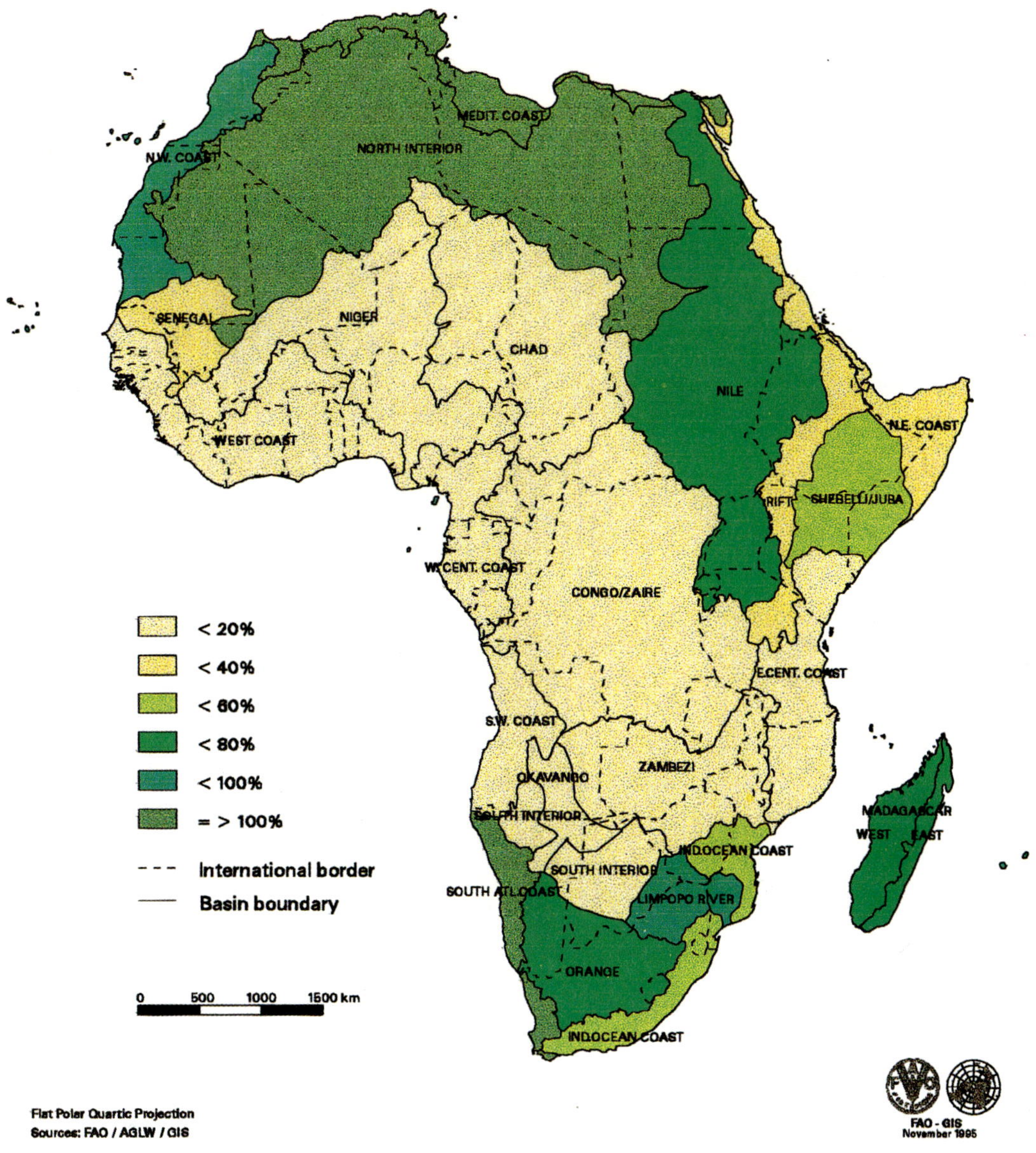

FIGURE 19
Possibilities for irrigation expansion (in millions of hectares per basin)

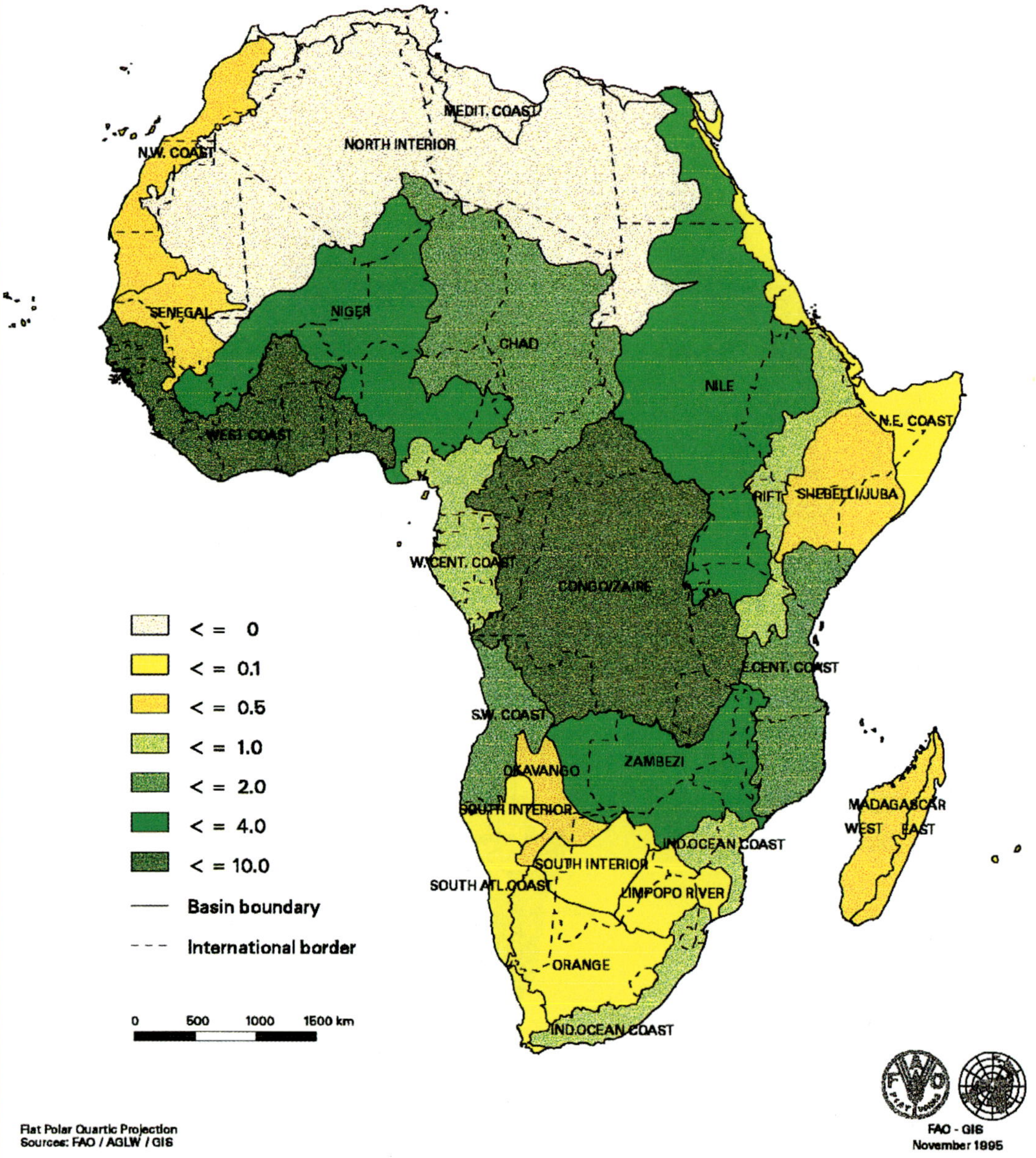

Flat Polar Quartic Projection
Sources: FAO / AGLW / GIS

FAO - GIS
November 1995

Chapter 9

Main sources of information

GENERAL

1. FAO/UNESCO. 1974. Soil map of the world, scale 1 : 5 000 000. FAO, Rome.

2. FAO/PNUD/IIASA [Institut International pour l'Analyse des Systèmes Appliqués]. 1984. Capacité potentielle de charge démographique des terres du monde en développement. Rapport technique du projet FPA/INT/513: Les ressources en terres des populations de demain. Préparé par Fischer, G., Higgins, G.M., Kassam, A.H., Naiken, L., Shah, M.M. 141 p. + 15 maps.

3. FAO. 1992. CROPWAT: A computer program for irrigation planning and management. FAO Irrigation and Drainage Paper 46. Rome. 133 p.

4. FAO. 1992. Crop water requirements. FAO Irrigation and Drainage Paper 24. Rome. 144 p.

5. FAO. 1995. FAO production yearbook 1994. Vol. 48. FAO Statistics Series No. 125. Rome. 243 p.

6. FAO. 1995. AGROSTAT PC diskettes on population, crop production and land use. Rome.

7. FAO. 1995. FAOCLIM cd-rom. Agroclimatic database. Rainfall and evaporation figures. FAO. Rome.

8. FAO. 1996. Food production: The critical role of water. Paper prepared for the World Food Summit, 11-17 November 1996. Advanced edition. Rome. 62 p.

9. Gleick, P. (ed). 1993. Water in crisis: a guide to the world's fresh-water resources. Oxford University Press for Pacific Institute. New York/Oxford. 497 p.

10. The International Journal on Hydropower and Dams. 1995. World atlas of hydroopower and dams. Volume 2, Issue 1, January 1995. 125 p.

11. WRI [World Resources Institute]/UNEP [United Nations Environment Programme]/UNDP [United National Development Programme]/WB [World Bank]. 1996. World Resources 1996-97. A guide to the global environment. The urban environment. Oxford University Press, New York and Oxford. 365 p.

AFRICA

12. BCEOM/MottMacDonald/ORSTOM/SOGREAH. 1992. Evaluation hydrologique de l'Afrique sub-Saharienne: Pays de l'Afrique de l'Ouest. Rapport préparé pour la Banque mondiale, le PNUD, la Banque africaine de développement et le Ministère français de coopération.

13. BRGM. 1991. Ressources en eau des pays africains, utilisations et problèmes. Préparé par Margat, J. Note technique 026 EAU/4S/91. BRGM, Services Sol et Sous-sols Département Eau. Orléans. 21 p.

14. Bulletin de l'Afrique Noire. 1976. L'économie des pays du Sahel. Numéro spécial du bulletin de l'Afrique noire. Ediafric. Paris. 154 p.

15. CILSS [Comité permanent Inter-Etats de Luttre Contre la Sécheresse dans le Sahel]/OCDE [Organisation de Coopération de Développement Economiques]. 1991. Le développement des cultures irriguées dans le Sahel. Rapport de synthèse et rapports par pays. OCDE/CILSS/Club du Sahel. SAH/D (91) 366. 216 p.

15a. Drijver, C.A. and Marchand, M. 1985. Taming the floods. Environmental aspects of foodplain development in Africa. Rijksuniversiteit Leiden. Netherlands.

16. FAO. 1978. Report on the agro-ecological zones project. Volume 1: Methodology and results for Africa. FAO, Rome.

17. FAO. 1986. Irrigation in Africa south of the Sahara. FAO Investment Centre Technical Paper 5. Rome. 182 p.

18. FAO. 1987. Consultation sur l'irrigation en Afrique. Bulletin FAO d'irrigation et de drainage no. 42. Rome. 221 p.

19. FAO. 1987. Irrigated areas in Africa: extent and distribution. FAO report AGL/MISC/13/87. Rome. 166 p. + map.

20. FAO. 1987. Irrigation and water resources potential for Africa. FAO report AGL/MISC/11/87. Rome. 127 p. + maps.

21. FAO. 1995. Water resources of African countries: a review. AGLW/FAO, Rome. 35 p.

21a. FAO. 1995. Irrigation in Africa in figures / L'irrigation en Afrique en chiffres. FAO Water Report 7. Rome. 336 p.

22. Godana, Bonaya Adhi. 1985. Africa's shared water resources: legal and institutional aspects of the Nile, Niger and Senegal River systems. A publication of the Graduate Institute of International Studies, Geneva. Frances Printer Publishers, London, Lynne Rienne Publishers, Colorado. 370 p.

23. Hutchinson, M.F., Nix, H.A., McMahon, J.P. and Ord, K.D. (in preparation). The development of a topographic and climate data base for Africa. Australian National University.

24. IFPRI. 1995. Water resources development in Africa: a review and synthesis of issues, potentials and strategies for the future. Report prepared for FAO by Rosegrant, M., Perez, W. and Nicostrato, D. International Food Policy Research Institute, Washington. 111 p.

25. OSS/UNESCO. 1995. Les ressources en eau des pays de l'OSS: évaluation, utilisation et gestion. Préparé par Margat, J. BRGM, Orléans. 83 p. + map..

26. Pérennès, Jean-Jacques. 1993. L'eau et les hommes au Maghreb: contribution à une politique de l'eau Méditerranée. Editions Karthala. Paris. 646 p.

27. UNESCO. 1972. Etude des ressources en eau du Sahara septentrional. Rapport final + 7 annexes techniques.

28. UNESCO. 1977. Atlas of world water balance: runoff coefficient map of Africa, scale 1 : 20 000 000. Paris. English translation of: Korzun, V.I. et al. 1974. Atlas of world water balance. USSR National Committee for the International Hydrological decade.

29. UNESCO. 1995. Discharge of selected rivers in Africa. Studies and reports in hydrology No. 52. 166 p.

30. United Nations. 1987. Groundwater in Northern and Western Africa. UN-DTCD Natural Resources Water Series No. 18. 415 p. + maps.

31. United Nations. 1988. Groundwater in Eastern, Central and Southern Africa. UN-DTCD Natural Resources Water Series No. 19. 320 p.

32. USAID Africa. 1987. African irrigation overview: main report. Prepared by J.R. Moris, D.J. Thorn and D.S. Humpal. Water Management Synthesis II Project. Utah State University, USA. 635 p.

RIVER BASINS

Senegal basin:

33. Park, Thomas K. (ed). 1993. Risk and tenure in arid lands: the political ecology of development in the Senegal River basin. University of Arizona Press, Tuscon & London. 408 p.

Niger basin:

34. CRWR [Center for Research in Water Resources]. 1995. Water balance of the Niger basin. Interim report prepared for FAO and UNESCO by the Center for Research in Water Resources, University of Texas at Austin. 54 p.

35. Lotti, C. & Associati. 1984-86. Reports of project FAO/UNDP DP/RAF/83/027: Assistance to the Niger Basin Authority. Inter-state Hydraulic Development of the Middle and Upper Reaches of the River Niger. Societa' di Ingegneria SpA., Rome.

36. ORSTOM. 1970. Monographie hydrologique du bassin du Niger. 3 volumes. Volume 1: Le Niger Supérieur et le Bani. 117 p. + annexes. Volume 2: La cuvette lacustre. 138 p. + annexes. Volume 3: Le Niger moyen. 113 p. + annexes.

Lake Chad basin:

37. Commission de bassin du Lac Tchad. Plan directeur pour le développement et la gestion écologiquement rationnelle des ressources naturelles du bassin conventionel du Lac Tchad. Préparé avec l'assistance et la collaboration du PNUE et de l'UNSO. 39 p. + 55 p. annexes.

38. DHV/SOGREAH. 1980. Lake Chad basin development study. Draft final report for UNDP.

39. Jauro, Abubakar B. 1995. The Lake Chad basin: problems and prospects. A paper presented at the ICID Conference, 11-15 September 1995, FAO, Rome. 23 p. + annexes.

40. UNDP/UNEP/UNSO. 1990. Le bassin conventionel du Lac Tchad: une étude diagnostique de la dégradation de l'environnement. Prepared by Arnould, E., Hutchinson, C., Kindler, J. and Warshall, P. 237 p.

Nile basin:

41. Howell, P.P. and Allan, J.A. (eds). 1994. The Nile: sharing a scarce resource: An historical and technical review of water management and of economical and legal issues. Cambridge University Press, Cambridge. 408 p.

42. Howell, P., Lock, M. and Cobb, S. (eds). 1988. The Jonglei Canal: Impact and opportunity. Cambridge University Press, Cambridge. 537 p.

43. KBO/UNDP. 1982. Development programme of the Kagera basin. Volume 1-6. Kagera Basin Organization and UNDP.

44. Said, Rushdi. 1993. The Nile River: geology, hydrology and utilization. Pergamon Press, Oxford, UK. 320 p.

45. SPIDER International Ltd. 1994. Water resources atlas of the Nile basin. Prepared for the Canadian International Development Agency.

46. UNDP. 1989. Nile basin integrated development. Fact finding mission draft report, projects RAF/86/003 and RAF/86/014.

Zambezi basin:

47. UNEP. 1987. Diagnostic study on the present state of ecology and the environmental management of the common Zambezi River system. UNEP/IG.78 Background paper 1. 101 p.

West Coast: Gambia basin:

48a. FAO. 1984. Agricultural research in the Gambia River basin: report of a review mission. Rome. 43 p.

West Coast: Volta basin:

48. ORSTOM. 1977. Le bassin du fleuve Volta: Monographies hydrologiques. Prepared by Moniod, F., Pouyaud, B. and Sechet, P. ORSTOM.

COUNTRIES

Algeria:

49. FAO. 1993. Projet d'appui à l'irrigation: rapport de préparation. 2 volumes. FAO Centre d'Investissement. Programme de coopération FAO/Banque mondiale. Rapport 67/93-CP-ALG 37. Rome.

50. Garadi, Ahmed. 1992. Prospective des besoins en eau et anticipation de la demande. Thèse pour le Doctorat de Sciences Economiques. Université de Grenoble. 269 p.

Angola:

51. FAO. 1980. Mission de formulation pour le secteur agricole. Rapport préparé par PNUD/FAO, Projet ANG/79/016. Deux volumes. Rome.

52. FAO. 1992. Mission d'identification générale de projets dans le secteur agricole. FAO Centre d'Investissement. Programme de coopération FAO/Banque mondiale/Banque africaine de développement. Rapport 3/92 CP/ADB-ANG 11. Rome.

53. MINADER. 1993. Plano Director de Irrigaçao em Angola. Ministério da agricultura e do desenvolvimento rural. 42 p.

54. Ministério da educaçao. 1982. Atlas geográfico, volume 1. Ministério da educaçao, República Popular de Angola. 49 p.

Benin:

55. Banque mondiale. 1992. Revue du secteur agricole. Rapport 10709-BEN. 58 p. + annexes + tableaux + cartes. Washington DC.

56. BCEOM/MottMacDonald/ORSTOM/SOGREAH. 1992. Evaluation hydrologique de l'Afrique sub-Saharienne: Pays de l'Afrique de l'Ouest. Rapport de pays: Bénin. Rapport préparé pour la Banque mondiale, le PNUD, la Banque africaine de développement et le Ministère français de coopération.

57. FAO. Dates diverses. Projets BEN/84/012 et BEN/91/002, Inventaire, étude et aménagement des bas-fonds. Rapports divers. Rome.

58. OMS. 1994. Développement agricole et santé au Bénin. Rapport d'un séminaire national, Cotonou, 23-26 novembre 1993. Rapport WHO/CWS/94.1. Genève. 66 p.

59. ORSTOM & Direction de l'hydraulique du Bénin. 1993. Ressources en eaux superficielles de la République du Bénin.

60. Piaton, H. 1986. Plan national d'irrigation en République populaire du Bénin. Rapport de la mission d'identification.

Botswana:

61. Department of Water Affairs. 1992. Botswana National Water Plan Study. Prepared by SMEC/KPB/SGAB. Ministry of Mineral Resources and Water Affairs.

62. FAO. 1991. Irrigation subsector review: prospects and constraints. Report prepared by Chapman, C. for project FAO TCP/BOT/0051: Assistance in irrigation development. Rome. 52 p.

63. Gieske, A., & Gould, J. (eds). 1994. Proceedings of the integrated water resources management workshop 1994. Kanye, Gaborone, 17-18 March 1994. Published by the Dept. of Environmental Science and Geology, University of Botswana, Gaborone, with SIDA funding. 214 p.

64. SADCC/AIDAB. 1992. Regional irrigation development strategy. Country report: Botswana. Harare. 60 p.

Burkina Faso:

65. Département de l'aménagement de la nature. 1994. Profil environnemental du Burkina Faso. Université agronomique. Wageningen.

66. Direction des statistiques générales. 1991. Annuaire statistique du Burkina Faso 1989-1990. Institut national de la statistique et de la démographie (INSD).

67. FAO. 1984. Haute Volta: Revue du sous-secteur de l'irrigation'. Annexe 4. FAO Centre d'Investissement. Programme de coopération FAO/Banque mondiale. Rapport 103/83 CP UPV 10.

68. FAO. 1989. Contribution à l'étude de l'inventaire de la mise en valeur et de l'aménagement des ressources en eau de surface et des terres irrigables du Sahel burkinabé (provinces Oudalan, Seno et Soum). Préparé par Edouard Buzingo. FAO, Rome. 132 p. + annexes.

69. FAO. 1989. Irrigation sector review desk study. FAO/World Bank Cooperative Programme. Report 28/89 CP-BKF 22 SR. Rome. 12 p. + annexes + maps.

70. FAO. 1993. Note de politique d'hydraulique agricole. Préparé pour le Ministère de l'eau et le Ministère de l'agriculture et des ressources animales, avec l'assistance de FAO dans le cadre du projet TCP/BKF/1353. Ouagadougou. 90 p.

71. FAO. 1995. Mission d'identification de projets d'investissement agricole. Annexe 1: Projet de valorisation des ressources en eau mobilisées par les petits barrages. FAO Centre d'Investissement. 43 p.

72. FAO. 1995. Projet de développement de l'irrigation privée'. FAO Centre d'Investissement. Rapport de pré-préparation. Rome.

73. IWACO et Direction des études et de la planification. 1991. Carte des ressources en eau de surface du Burkina Faso. Ministère de l'eau.

74. Ministère de l'eau et Ministère de la coopération des Pays-Bas. 1991. Etude du bilan d'eau au Burkina Faso. Inventaire des ressources en eau. Version définitive. Tome II: Rapports régionaux. Ouagadougou.

75. Ministère de l'eau et Ministère de la coopération des Pays-Bas. 1991. Etude du bilan d'eau au Burkina Faso. Etude du schéma directeur d'approvisionnement en eau potable du Burkina Faso (1990-2005). Version définitive. Tome I: Rapport national. Ouagadougou.

76. ONBAH. 1987. Inventaire et reconnaissance générale de l'état des barrages et retenues d'eau au Burkina Faso. Ministère de l'eau.

Burundi:

77. World Bank. 1993. Private sector development in agriculture. Report 9686-BU. Washington.

78. PNUD/FAO. 1985. La mise en valeur hydro-agricole au Burundi: Etat actuel et stratégie pour l'avenir. Préparé par van Leeuwen, N. Rome.

Cameroon:

79. FAO. 1990. Mission d'identification générale de projets agricoles. FAO/DDC Rapport 103/90 AF-CMR 29. Rome.

80. United States Department of Agriculture et Fonds d'Aide et de Coopération de la France. 1977. Inventaire des ressources du Nord du Cameroun, Afrique. 188 p.

Cape Verde:

81. CILSS. 1989. Etude sur l'amélioration des cultures irriguées au Cap-Vert. Rapport de synthèse. Ougadougou.

82. PNUD/CNAG. 1993. Schéma directeur pour la mise en valeur des ressources en eau (1993-2005). Rapport préparé par le projet CVI/81/001: Assistance à la JRH.

Central African Republic:

83. Direction météorologie nationale. Annuaires hydrologique 1990/91, 1991/92, 1992/93, 1993/94. Préparé par le projet CAF/91/021. Ministère des transports, des travaux publics, de l'habitat et de l'aménagement du territoire.

84. ORSTOM. 1987. Carte oro-hydrographique de la Républic Centrafricaine. Préparé par Boulvert, Y.

Chad:

85. FAO. 1987. Schéma directeur pour la mise en valeur des petit aménagements hydro-agricoles. Préparé par Nisse, M., De Nooy, E.J.P. Projet FAO TCP/CHD/6754. Rome. 61 p. + annexes.

86. FAO. 1989. Irrigation subsector review. FAO Investment Center. FAO/World Bank cooperative programme. Report 37/89 CP-CHD 11 SR. Rome. 26 p. + annexes.

87. FAO. 1994. Programme d'Action National. Rome. 67 p.

Comoros:

88. BDPA. 1991. Etude de la stratégie agricole des Comores.

89. Ministère du développement rural, de la pêche et de l'environnement. 1994. Consultation sectorielle sur l'environnement et l'agriculture.

Congo:

90. BCEOM/MottMacDonald/ORSTOM/SOGREAH. 1992. Evaluation hydrologique de l'Afrique sub-Saharienne: Pays de l'Afrique de l'Ouest. Rapport de pays: Congo. Rapport préparé pour la Banque mondiale, le PNUD, la Banque africaine de développement et le Ministère français de coopération.

Côte d'Ivoire:

91. BCEOM/MottMacDonald/ORSTOM/SOGREAH. 1992. Evaluation hydrologique de l'Afrique sub-Saharienne: Pays de l'Afrique de l'Ouest. Rapport de pays: Côte d'Ivoire. Rapport préparé pour la Banque mondiale, le PNUD, la Banque africaine de développement et le Ministère français de coopération.

Djibouti:

92. Bundesanstalt für Geowissenschaften und Rohstoffe. 1982. Inventaire et mise en valeur des ressources en eau de la République de Djibouti. Rapport préparé par Müller, W. pour le projet No. 78.2233.1, Coopération hydrogéologique allemande. Hannover, Germany. 176 p.

93. Direction nationale de la statistique. 1991. Annuaire statistique de Djibouti. Résultats de 1990. Ministère du commerce, des transports et du tourisme.

Egypt:

94. CAPMAS. 1993. Description of Egypt.

95. Egyptian Environmental Affairs Agency. 1992. Environmental action plan of Egypt.

96. FAO. 1993. National Action Programme. Rome. 38 p.

97. Nasser Ezzat, M. 1994. Water resources development. Country report: Egypt. Prepared for the Nile 2002 Conference. 31 p.

98. Othman, Y. 1994. Experiences in integrated land and water management. In: Proceedings of the ISAWIP final seminar. Port-Said, Egypt, 18-21 April 1994.

Equatorial Guinea:

99. FAO. 1994. Natural resources management and conservation project. Preparation report. FAO/DDC report 12/94 CP-EQG11. Rome.

Eritrea:

100. FAO. 1994. Agricultural sector review and project identification. 3 volumes. FAO/IC project TCP/ERI/2353. Rome.

Ethiopia:

101. Aytenffisu, M. 1981. Groundwater in Ethiopia. DTCD/ECA: Groundwater in Africa. 22 p.

102. FAO. 1990. Development of irrigated agriculture, phase 2. Terminal Report: project findings and recommendations. Project ETH/82/008. Rome. 43 p.

103. FAO. 1990. Irrigation policy and strategy in Ethiopia. Project TCP/ETH/8963.

104. FAO. 1994. Small-scale irrigation consolidation project. Preparation report. FAO/DDC Report 59/94 ADB-ETH 48. 58 p. + maps + figures + tables + appendixes + annexes.

105. FAO. 1994. Studies for integrated irrigation systems: project findings and recommendations. Terminal Report project UNDP/FAO ETH/88/001. Rome. 55 p.

106. FAO. 1995. A note on small-scale irrigation in Ethiopia. Rome. 10 p.

107. The Nile 2002 Conference. Water supply and demand in the Ethiopian Nile sub-basin. February 13-17, 1994. Arusha, Tanzania. 12 p.

108. UNDP. 1993. Fifth cycle country programme. Sub-programme 2B: Utilisation of natural resources for balanced development. Component 2: water resources development. First draft. Addis Abeba.

Gabon:

109. FAO. 1988. Mission de reconnaissance/identification de projets agricoles FAO/BAD. Programme de soutien à l'investissement. FAO Centre d'investissement, Rapport 90/88 AF GAB 7 WP. Rome. 22 p.

The Gambia:

110. FAO. 1986. Review of short and medium-term irrigation policies. Terminal statement of project TCP/GAM/4504. Rome.

111. FAO. 1994. Lowlands agricultural development programme. FAO Investment Centre. FAO/IFAD Cooperation Programme report 52/94 IFAD-GAM 15. Rome.

Ghana:

112. Water Resources Research Institute/FAO. 1993. Survey on water use for agricultural and rural development. 59 p.

113. World Bank. 1986. Irrigation subsector review. Report 6173-GH. 60 p. + maps.

114. World Bank. 1990. Medium-term agricultural development programme. Washington DC.

Guinea:

115. Direction nationale du génie rural. 1991. Lettre de politique de développement agricole. Note thématique nr. 12: Aménagements hydro-agricoles. Ministère de l'agriculture et des ressources animales.

116. FAO. 1995. Développement de la petite irrigation en Guinée. FAO/AGLW, Rome. 11 p.

Guinea-Bissau:

117. FAO. Dates diverses. Enquête annuelle sur les superficies, rendements et productions Résultats pour les campagnes 1990/91, 1991/92, 1993/94. Projet PNUD/FAO GBS/90/004. Rome.

118. PNUD. 1991. Rapport du projet GBS/87/002: Schéma directeur pour le secteur eau et assainissement. 98 p. + annexes.

119. PNUD/BAD/Ministère français de la coopération. 1992. Evaluation hydrologique de l'Afrique sub-Saharienne, Pays de l'Afrique de l'Ouest. Rapport de pays: Guinée Bissau. Projet PNUD RAF/87/030.

120. PNUD/FAO. 1993. Bilan diagnostic du secteur agricole. Projet GBS/92/T01/A.

Kenya:

121. Central Bureau of Statistics. 1992. Statistical abstract 1991.

122. Euroconsult/Delft Hydraulics Laboratory/Royal Tropical Institute. 1987. Study on options and investment priorities in irrigation development.

123. FAO. 1985. A review of small-scale irrigation schemes in Kenya. Prepared by van Doorne, J.H. FAO Report AGL/MISC/2/85. Rome. 38 p. + annexes.

124. Irrigation and Drainage Branch. 1990. Atlas of irrigation and drainage in Kenya. Small-scale Irrigation Development Project (SIDP). Ministry of Agriculture. 132 p.

125. Ministry of Water Development. 1992. The study on the national water master plan. Prepared with the assistance of Japan International Cooperation Agency (JICA).

126. Ministry of Water Development. 1985. Water requirements for irrigation in Kenya. Prepared by Mesny, M. and Kalders, J.

Lesotho:

127. Ministry of Agriculture. 1991. The history of irrigation in Lesotho. Prepared by Mohapeloa, Z.K.M. In proceedings: Irrigation in Lesotho.

128. Ministry of Natural Resources. 1994. Hydrogeological map of Lesotho.

Liberia:

129. FAO. 1986. Report of an agricultural sector review mission. Report 85/86 CP-LIR.8. 32 p. + appendixes + annexes.

Libya:

130. Pallas, P. 1980. Water resources of the Socialist People's Libyan Arab Jamahiriya. In: The Geology of Libya. Proceedings of the second symposium on the geology of Libya. London Academic Press. Pp. 539-594.

131. Salem, O.M. 1992. The great man-made river project. Water Resources Development, Volume 3, Number 4, December 1992. Pp. 270-278.

Madagascar:

132. Banque mondiale. 1992. Etude sectorielle de l'agriculture irriguée. Division agriculture, Département des pays des grands lacs et de l'océan Indien, Région Afrique. Washington DC.

133. Boumendil, Henry. 1991. Contribution à la revue du sous-secteur de l'irrigation.

134. FAO. 1993. Projet de réhabilitation des petits périmètres irrigués. FAO/CI rapport 9/93 CP-MAG 35. Rome. 2 volumes.

Malawi:

135. SADCC. 1992. Regional irrigation development strategy. Country report: Malawi.

136. UNDP. 1986. National water resources master plan. Projects MLW 79/015 and MLW 84/003. Department of Water, Ministry of Works and Supplies.

137. Water Department. 1994. Integrated water resources management plan for Zambezi River Basin. Country position paper prepared by Chirwa, A.B. for the Workshop at Livingstone, Zambia, 2-6 May 1994. Ministry of Works. 33 p.

Mali:

138. FAO. 1992. Aperçu sur le sous-secteur irriguée. FAO Centre d'Ivestissement. Rapport 84/92 CP-MLI 36 SR. 39 p. + cartes.

139. Ministère de l'agriculture, République du Mali. 1987. Revue du secteur agricole du Mali. 345 p.

140. PNUD. 1990. Schéma directeur de la mise en valeur des ressources en eau du Mali. Projet PNUD/UNDTCD MLI/84/005.

141. World Bank/UNDP. 1985. Study on options and investment priorities in irrigation development. Project INT/82/001. 192 p.

Mauritania:

142. FAO. 1986. Projet de désenclavement et d'aménagement de moyens périmètres: rapport de préparation. FAO Centre d'Investissement. Programme de coopération FAO/Banque Africaine de Développement. Rapport 76/86 AF-MAU 16.

143. FAO. 1988. Projet d'amélioration de l'irrigation: rapport de préparation. FAO Centre d'Investissement. Programme de coopération FAO/Banque mondiale. Rapport 1/88 CP-MAU 21. 2 volumes.

144. FAO. 1995. Note on small irrigation. AGLW/FAO. Rome.

145. Ministère du développement rural et de l'environnement, République Islamique de Mauritanie. 1993. Annuaire statistique des oasis. Projet Oasis. 71 p.

Mauritius:

146. World Bank. 1992. Mauritius: expanding horizons. Report 9685-MAS. Washington DC.

Morocco:

147. Amicale des Ingénieurs du Génie Rural (AIGR). 1991. Annuaire 1990 du génie rural.

148. Association nationale des améliorations foncières, de l'irrigation et du drainage (ANAFID). 1990. L'irrigation au Maroc.

149. Banque mondiale. 1995. Le secteur de l'eau au Maroc. Rapport 12649-MOR. 65 p.

150. Conseil supérieur de l'eau. 1993. Aménagement hydro-agricole: situation actuelle et perspectives. 7ème session. 43 p. + cartes.

151. Conseil supérieur de l'eau. 1993. Plan directeur intégré d'aménagement des eaux des bassins du Loukkos, du Tangerois et des cotiers méditerranéens. 7ème session. 122 p.

152. Lahlou, O. 1988. L'aménagement hydro-agricole: cas du périmètre irrigué marocain.

153. Office national de l'eau potable. 1993. Le secteur de l'eau potable: bilan besoins-ressources. Direction Planification et Développement, Division Dégogenes de Ressources. 7 p.

154. Rieul, L. 1994. Le programme national d'irrigation du Royaume du Maroc et la coopération Franco-marocaine. Dans: Bulletin du Conseil Général du GREF, numéro 40, décembre 1994.

Mozambique:

155. Direcçao nacional de aguas. 1995. Proposals for a national water policy. Preliminary version. Ministerio de Construcçao e Aguas. 64 p.

156. Ferro, B.P., Bouman, D. 1987. Hydrogeological map of Mozambique. Ministry of Construction and Water/UNICEF. 46 p.

157. Ministerio da Educaçao. 1986. Atlas Geográfico. Volume 1. Républica Popular de Moçambique. 49 p.

158. SOGREAH-HIDROGEST. 1993. National irrigation development master plan. Final report.

159. UNESCO/UNDP. 1984. Present situation of water resources management in Mozambique. Project UNDO/UNESCO MOZ/81/001. National Water Directorate. Maputo.

Namibia:

160. Department of Water Affairs, Ministry of Agriculture, Water and Rural Development, Republie of Namibia. 1993. Central area water master plan: phase 1. Executive summary. GTZ.

161. Ministry of Agriculture, Water and Rural Development. 1994. Identification and prioritization of irrigation development opportunities on the Orange River. Report No. 2010-03.

162. National Planning Commission. 1993. Transitional national development plan 1991/92-1993/94. Windhoek.

163. SADCC. 1992. Regional irrigation development strategy. Country report: Namibia. Harare. 72 p.

164. UNDP. 1989. Base studies on financial, economic and social aspects for the arrangements for independence in Namibia: assessment of the water resources sector. 50 p.

Niger:

165. FAO. 1989. Note on irrigation. FAO/World Bank cooperative programme. Report 83/89 CP-NER 16 DS. 8 p. + tables + maps + annexes.

166. FAO. 1991. Projet de promotion de la petite irrigation privée. Mission de préparation. FAO Centre d'Investissment. Programme de coopération FAO/Banque mondiale. Rapport 42/91 CP-NER 23. 77 p. + annexes.

167. FAO. 1993. Schéma directeur de mise en valeur et de gestion des ressources en eau. Projet NER/92/007. 105 p. + annexes.

168. Seini Ali. 1990. Les eaux de surface et leur utilisation actuelle ou envisagée. Consultant du projet PNUD/DCTD NER/86/001.

169. SOGREAH/BRGM. 1981. Etude du plan de développement de l'utilisation des ressources en eau du Niger.

Nigeria:

170. Adams, W.M., Aminu-Kano, M., Hollis, G.E. 1993. The Hadejia-Nguru wetlands: environment and sustainable development of a Sahelian floodplain wetland. IUCN, Gland, Switzerland. 244 p.

171. FAO. 1992. Irrigation subsector review. FAO Investment Centre. Report 89/91 CJP-NIR 45 SR. 61 p. + annexes.

172. Japan International Cooperation Agency (JICA). 1993. The study on the national water resources master plan. Federal Ministry of Agriculture, Water Resources and Rural Development.

173. MottMacDonald International. 1992. Sub-Saharan Africa hydrological assessment. West African Countries: Country report: Nigeria. Report prepared with the assistance of BCEOM and SOGREAH for the World Bank/UNDP/ADB.

Rwanda:

174. Equipe interdisciplinaire intégrée de terrain/AID-WMSII. 1987. Rapport sur la stratégie nationale pour le développement et la gestion des petits marais.

175. FAO. 1991. Rapport de mission en hydrology au projet PNUD/FAO RWA/89/006. Préparé par Szkutnicki, J.

176. FAO. 1993. Inventaire des marais du Rwanda: conclusions et recommandations. Compte rendu final du projet PNUD/FAO RWA/89/006.

177. Prioul, C. 1981. Atlas du Rwanda.

São Tomé and Principe:

178. FAO. 1981. Formulation d'un projet sur l'inventaire des ressources hydraulique à São Tomé et Principe. Projet TCP/STP/0103. Rome.

Senegal:

179. FAO. 1985. Politiques et programmes céréaliers. Une analyse du secteur céréalier et un programme d'actions pour son développement.

180. FAO. 1993. Mission d'identification d'un projet de développement de la filière rizicole dans la vallée du fleuve Sénégal. Rapport de synthèse.

181. Ministère de l'hydraulique/PNUD. 1994. Bilan-diagnostic des ressources en eau du Sénégal. Préparé par le projet MH/PNUD/DADSG-SEN/87/006: Planification des ressources en eau.

182. Raes, D., Sy, B., Van Passel, L. 1992. The water balance of rice irrigation schemes in the Senegal River delta. Irrigation water management project. In: Advances in Planning, Design and Management of Irrigation Systems as related to Sustainable Land Use. Center for Irrigation Engineering, KU Leuven, Belgium. Volume 2, pp. 835-844.

182a. World Bank. 1987. Irrigation IV project: Staff appraisal report. Report 6484-SE. Washington DC.

Seychelles:

183. FAO. 1989. La stratégie du secteur agricole. FAO Centre d'Investissment. Rapport 88/89 TA-SEY 4. Rome. 39 p.

Sierra Leone:

184. FAO. 1976. Rice development project: preparation mission. FAO Investment Centre. Report DDC 15/76 SIL.3. Rome.

185. FAO. 1981. Land resources survey project DP/SIL/73/002. Project Finadings and Recommendations. Terminal report. Rome. 75 p.

Somalia:

186. FAO. 1989. A brief description of major drainage basins affecting Somalia. National Water Centre, Mogadishu. Field Document No. 14 prepared by Kammer, D. Project FAO SOM/85/008. Rome. 30 p. + maps.

187. Indian Ocean Newsletters. 1985. Productive sectors of the economy.

188. World Bank. 1987. Agricultural sector survey: main report and strategy. Report 6131-SO. Washington DC. 86 p. + annexes.

South Africa:

189. Department of Agriculture and Department of Water Affairs and Forestry. 1994. Irrigation strategy. 14 p. + tables.

190. Department of Water Affairs. 1986. Management of the water resources of the Republic of South Africa.

191. FAO. 1992. Agricultural sector mission: Irrigation reconnaissance. FAO Investment Centre. FAO/World Bank cooperative programme. Report 123/92 CP-RSA 1 SR/WPs.

192. Water Research Commission. 1994. Irrigation development in southern Africa with special reference to South Africa. Prepared by van der Merwe, David and Reid, Peter. Pretoria. 19 p.

Sudan:

193. Abdelsalam, A. 1991. Sedimentation in Sudan multipurpose reservoirs. International Publ. Hydraulic Research Station-Sudan. 8 p.

194. Adam, Ahmed M. *et al.* 1995. The Republic of Sudan: cooperative environmentally sound and integrated development of the Nile basin. Proceedings Nile 2002 Conference, February 1-6, 1995.

195. Craig, G.M. (ed). 1991. The agriculture of the Sudan. Oxford University Press, Oxford, UK. 468 p.

196. ICID. 1995. Comments on draft country profile for FAO Water Reports 7: Irrigation in Africa in figures.

Swaziland:

197. Department of the Army, US Corps of Engineers. 1981. Water and related land resources framework plan.

198. World Bank/UNDP. 1990. Sub Saharan Africa hydrological assessment. Country report: Swaziland.

Tanzania:

199. FAO. 1990. Smallholder irrigation development priorities. Field document No. 14 prepared by Chapman, C. Project FAO/UNDP URT/86/012: Institutional support for irrigation development. Rome. 49 p. + annexes.

200. FAO. 1993. Draft national irrigation development plan. Project UNDP/FAO URT/90/016. Rome. 89 p.

201. FAO. 1993. National Action Programme. FAO/IAP-WASAD. Rome. 77 p.

202. Kilimo. 1984. Irrigation development in Tanzania. 18 p.

Togo:

203. FAO. 1991. Revue du secteur agricole. FAO Centre d'Investissment. Programme de coopération FAO/Banque mondiale. Rapport 45/91 CP-TOG 13 SR.

Tunisia:

204. Direction générale du génie rural. 1992. Rapport du Comité Irrigation - 8ème Plan. Ministère de l'agriculture.

205. Direction générale du génie rural. 1994. La gestion de l'eau en Tunisie. Ministère de l'agriculture.

206. Direction générale des ressources en eau. 1992. Commission de réflexion sur le développement des ressources en eau de surface. Ministère de l'agriculture. 41 p.

207. World Bank. 1985. Irrigation management improvement project. Staff appraisal report 5396-TUN. Washington DC. 50 p. + tables + charts + annexes.

208. World Bank. 1994. Water sector review. Report 13054-TUN. Washington DC. 32 p. + annexes.

Uganda:

209. Statistisches Bundesamt. 1991. Country profile Uganda.

210. Wasawo, David P.S. (ed). Strategic Resources Planning in Uganda. Volume IV: water resources. UNEP.

Zaire:

211. Banque mondiale. 1988. Memorandum sur le secteur agricole. Zaire: vers un développement agricole soutenu. Rapport 7356-ZR. Washington DC. 64 p. + maps.

Zambia:

212. JICA. 1994. The study on the national water resources master plan in the Republic of Zambia. Progress Report No. 2 prepared by Yachingo Engineering Co. Ltd.

213. Ministry of Agriculture, Food and Fisheries. 1994. Agricultural sector investment programme: irrigation sub-programme. Prepared by Akayombokwa, I.M., Chibinga, P., Choseni, P., Moono, D. 46 p.

214. SADC. 1992. Regional irrigation development strategy. Country report: Zambia. Harare.

215. World Bank/Kingdom of the Netherlands. 1989. Study on options and investment priorities in irrigation development. Report prepared by Euroconsult/Delft Hydraulics Laboratory/Royal Tropical Institute.

Zimbabwe:

216. FAO. 1990. Irrigation subsector review and development strategy. Project TCP/ZIM/8955. Rome. 55 p. + 6 technical annexes.

217. FAO. 1993. Country Action Programme: Zimbabwe. 92 p.

FAO LAND AND WATER BULLETINS

1. Land and water integration and river basin management, 1995 (E)
2. Planning for sustainable use of land resources – Towards a new approach, 1995 (E)
3. Water sector policy review and strategy formulation – A general framework, 1995 (E)
4. Irrigation potential in Africa – A basin approach, 1997 (E)

Availability: April 1997

Ar – Arabic		Multil –	Multilingual
C – Chinese		*	Out of print
E – English		**	In preparation
F – French			
P – Portuguese			
S – Spanish			

The FAO Technical Papers are available through the authorized FAO Sales Agents or directly from Sales and Marketing Group, FAO, Viale delle Terme di Caracalla, 00100 Rome, Italy.